Laser Space Communications

For a complete listing of the *Artech House Space Technology and Applications Series,* turn to the back of this book.

Laser Space Communications

David G. Aviv

ARTECH
HOUSE

BOSTON | LONDON
artechhouse.com

A catalog record for this book is available from the U.S. Library of Congress.

A catalogue record for this book is available from the British Library.

ISBN 1-59693-028-4
ISBN 978-1-59693-028-5

Cover design by Robert Pike.

© 2006 by David G. Aviv
Published by ARTECH HOUSE, Inc.
685 Canton Street
Norword, MA 02062

10 9 8 7 6 5 4 3 2 1

The author wishes to express his deepest appreciation to his three sons and his wife, who inspired the writing of this book and contributed to its completion.

Contents

	Preface	*xv*

1	**Introduction**	**1**
1.1	Overview	1
1.2	Advantages of Lasers over Microwaves	2
1.2.1	Narrow Beamwidth	2
1.2.2	Large Directivity	3
1.2.3	Higher Bandwidth of Lasers Versus Microwaves	4
1.2.4	Privacy Comparison Between Laser and Microwave	4
1.3	Combined Laser and Microwave Communication	5
1.4	Combined Signal-to-Noise for Microwaves in Uplink/Downlinks and Laser Crosslinks	7
1.5	Review of the Chapters	8
1.5.1	The Essence of Chapter 2	8
1.5.2	The Essence of Chapter 3	9
1.5.3	Overview of Chapters 4, 5, and 6	10
1.5.4	The Essence of Chapter 7	12
1.5.5	The Essence of Chapter 8	13
1.5.6	The Essence of Chapter 9	14

2	**The Signal Power Budget for Intersatellite Links and Potential Mars-to-Earth Links**	**15**
2.1	Introduction	15
2.2	Signal Power Budget Calculation	16
2.2.1	Numerical Example	18
2.3	Summary of the Power Budget Expression	20
2.4	Evaluation of BER as a Function of Photoelectrons per Bit and Modulation Scheme	21
2.4.1	Error Analysis for PGBM	22
2.4.2	Example of Noise Background Calculations	24
2.4.3	Detailed Background Calculations	24
2.5	Direct Detection Versus Heterodyne Detection	25
2.5.1	Signal-to-Noise Ratio for the Direct Detection Receiver	26
2.5.2	Signal-to-Noise Ratio of the Heterodyne (Coherent) Receiver	28
2.5.3	Other Modulation Formats	29
2.5.4	Laser Communication Between Mars and Earth Using PPM Modulation	34
2.6	Expression of Signal Power Budget Due to Vibrations	35
2.6.1	The Pointing Loss Factors	36
2.6.2	Mathematical Expressions for the Pointing Losses	36
2.6.3	Satellite Vibrations and Their Effect on the Communication Link Performance for a Typical Laser Transceiver Design for ISL	37
2.6.4	Effect of Vibration on the Communication Links Through a Constellation of Satellites	38
2.7	Azimuthal and Elevation Components of the Pointing Error Angle in a Constellation of Satellites	39
2.8	Summary and Concluding Remarks	42
	References	42

3   Acquisition Tracking and Pointing   45

3.1   Introduction   45

3.2   Implementation of the ATP Functions   47
3.2.1 Functional Description   48

3.3   Basic Block Diagram of the ATO on LEO
      and GEO   51

3.4   Specific Acquisition Procedures in Step 1   54

3.5   Step 2 of the Acquisition Process   61

3.6   Step 3 of the ATP Process   61
3.6.1 Use of the GPS to Determine the Location
      of the LEO   62
3.6.2 Acquisition Timing   63
3.6.3 Acquisition Timing Using GPS   63

3.7   Additional Tracking Considerations   64

3.8   Integration of the ATP Within the
      Laser Transceiver   66
3.8.1 The Laser Transceiver for the LEO Satellite   66
3.8.2 Baseline Laser Transceiver with Inertial Sensors   68
3.9   Summary and Concluding Remarks   71

      References   72

4   Satellite Downlink Through the Atmosphere   73

4.1   Introduction   73

4.2   Downlink from Satellite to Ground Station   75

4.3   Analytic Expressions of the Downlink Signal   77

4.4   Variation of Angle of Arrival of the
      Downlink Signal   82

4.5   Summary and Concluding Remarks   83

      References   84

**5   Uplink Laser Communication Through
        the Atmosphere   87**

5.1   Introduction   87

5.2   Differences Between Downlink and Uplink   88

5.3   Calculating Signal Coupling Efficiency   89

5.4   Coherence Length and Associated
        Atmospheric Turbulence   92

5.5   Measurement of Atmospheric Effects
        on Downlink and Uplink   93

5.6   Methods of Obtaining a "Reference"
        Downlink Signal for Adaptive
        Optics Subsystem   95
5.6.1   Synthetic Sodium Laser Beacon   96

5.7   Using a Reference Laser and an Oriented
        Mirror   97

5.8   Uplink Signal Loss When AOS Is Not Used   98

5.9   Summary and Concluding Remarks   101

        References   101

**6   Terrestrial Laser Communication Links and
        Weather Issues   103**

6.1   Introduction   103

6.2   Calculations of Atmospheric
        Turbulence Parameters   104

6.3   Absorption and Scattering in the Atmosphere   105

6.4   Attenuation Due to a Variety of
        Weather Components   111
6.4.1   MODTRAN System for Estimating Laser
        Signal Penetration of the Atmosphere   112

6.5   The Weather Avoidance System   113

6.5.1   Examples of Dry Weather Locations in the Southwest Region of the United States   113

6.5.2   Pictorial Representation of the Weather Avoidance System   118

6.6   Testing of Laser Communications Along Terrestrial Links   118

6.7   Concepts for Penetrating Low Clouds Surrounding a Ground Station or Airstrip   120

6.8   Summary and Concluding Remarks   121

    References   122

7   The Fifth-Generation Internet System   123

7.1   Introduction   123

7.2   The Synchronous Laser Backbone   124

7.2.1   Weather Effects   126

7.3   Example of Weather Satellite for 5-GENIN   127

7.4   Unique Requirements of Atmospheric and Earth-Based Laser Terminals   129

7.4.1   Protection of Stationary, High-Value Targets   130

7.4.2   Observation and Monitoring of Borders   131

7.4.3   Airship Vulnerability Issues   132

7.4.4   Endurance of Aerostate After Suffering Enemy Fire   133

7.5   The Miniaturized Unmanned Ground-Based Mobile Systems   134

7.5.1   Unique Applications of the MUGMs   135

7.6   Ground-Based Power Support for the Backbone Satellites   135

7.6.1   Additional Features of the 5-GENIN System   136

7.7   Summary and Concluding Remarks   138

    References   139

8   Passive Reflector Configurations   141

8.1   Introduction   141
8.1.1   History of Passive Reflectors, the RF Case   145
8.1.2   Extension of Applications in the Passive Field   146

8.2   The Nominal Reference Link   147
8.2.1   Data Rate, Modulation Scheme, and Range   148
8.2.2   A Simplified Signal Power Budget for a
           Reflective Structure   149

8.3   Selected Passive Reflectors   150
8.3.1   The Articulating Mirror System   150
8.3.2   Calculations of Data Rate for Articulating
           Mirror System   150

8.4   Experiments Using Passive Spatial Reflectors   153

8.5   High-Energy Deposition   156

8.6   The Optical Westford   158

8.7   Additional Design Considerations   159
8.7.1   Use of GPS, Ground, and Retrodirective
           Mirrors to Locate Reflective Faces   159
8.7.2   Trades: Increasing the Signal Bandwidth   160

8.8   Summary and Concluding Remarks   160

           References   161

           Appendix 8A   163

9   Unique Applications of Laser Communications   165

9.1   Introduction   165

9.2   Combined RF and Laser Telescope Antenna   167

9.3   Space-Based ASW to Achieve Detectability
           and Identification of Submersibles   168
9.3.1   Blue-Green Laser System Design Features   169

9.3.2   Experimental Approach to Achieve the
          SBDIS Lidar   169

9.4   Submarine Laser Communication to
          Satellite Concept   171
9.4.1   Loss Due to Beam Spread Resulting from
          Seawater Turbulence   173
9.4.2   Beam Spread Due to Suspended
          Biological Particles   174

9.5   Approximation of Laser Signal Loss in Satellite-
          to-Submarine Communication   177

9.6   Return Link Communication   177

9.7   Interplanetary Laser Communication   178
9.7.1   Feasibility of Laser Communications Between
          Mars and Earth   179
9.7.2   Further on the Mars-to-Earth Laser
          Communication Potential   180

9.8   Proposed Laser Communication Between the
          Moon and Earth   181

9.9   The Microsatellite or Nanosatellite   181

9.10   The SALT System   183

9.11   The Retroreflective Communication System   185

9.12   Summary and Concluding Remarks   187

          References   188

          About the Author   **189**

          Index   **191**

Preface

In the next two decades the technology of laser communications is very likely to be improved and augmented for space applications as well as airborne and ground-based platforms. This will enable the development of integrated equipment at all three levels and also broaden its utility with extreme broadband signals contained in narrow beam links throughout the world's communication networks.

Laser technology provides privacy and interconnectivity with potential Internet in space with little power demand, compact size, and low weight. Not requiring governmental frequency assignment is another advantage of laser communications.

In this book a number of new approaches are discussed in the area of laser communication applications. Included in the communication network is the fifth generation Internet (5-GENIN), synchronous altitude backbone (SAB), weather avoidance system (WAS), tactical communications covering fixed- and rotary-wind aircraft, ships at sea, ground-based stations both fixed and moving, and a variety of airships designed to monitor land and sea borders and high-value targets both military and civilian.

Downlink and uplink through the atmosphere are considered, with particular attention toward mitigation of the effect of atmospheric turbulence. Finally, it is demonstrated by calculations that various links are feasible between artificial statellites and Earth, between the Moon and Earth, and between Mars and Earth. Although there is not at the moment a funding commitment for such laser links, the knowledge and technological base to implement these communications is substantial.

In terms of general communications, the ability to provide very wide bandwidth capacity in the visible and near wave infrared, combined with RF communications, should enable the functions of observation, monitoring, and command; control communications will be achieved through the transformational communication satellite system. The linking of every element of our interest, including unmanned airborne vehicles (UAVs), miniaturized unmanned ground-based mobile (MUGM) system and other ground and airborne vehicles, are all touched on in this book.

Over more than thirty years of work in laser space communications, I have learned from and interacted with many physicists and electronics systems and communications engineers. In particular I wish to thank Hal Yura, Art Kraemer, Jerrry Gelbwachs, Hal Stoll, Arnie Silver, Tom Hartwick, Max Weiss, Abhijit Biswas, Monte Ross, Harris Rawicz, Steve Feldman, Renny Fields, Lenny Bergstein, Arnold Newton, Stan Sadin, Ted Taylor, Joe Statsinger, Joel Anspach, and Schlomi Arnon. Their work and insights have shaped the courses have I taught in this field. They are a special inspiration to me and to my students.

Finally, I wish to acknowledge the real contributions of my dear family to the writing of this book. Oren encouraged the writing of *Laser Space Communications* and, together with his wife Katie, often advised and resolved various issues in the preparation of the text; Dr. Jonathan helped me to establish the appropriate breadth of the technical material; and Bobby designed the computer station and its interconnectivity. Lastly, my BW Rena saw hundreds of pages strewn around the house from the entrance door to my office, via the living room, dining room, bedroom, and coffee station, and never complained. With all seriousness, it would be correct to say that my five family members all contributed with their unique engineering talents to this effort—whether or not they themselves recognized it. I thank them from the bottom of my heart.

1

Introduction

1.1 Overview

This book describes the engineering aspects of laser space communications systems. It will enable the electrical engineer to design laser data links in a variety of environments.

Having taught laser communications for a number of years, the author believes that this laser space communications book will be a useful tool for the advanced undergraduate student, the graduate student, and the practicing engineer in industry and government. It will help in the design of laser links in space and in the atmosphere, covering the bandwidths, error rate, privacy aspects, platform stability, adaptive optical subsystems, weather avoidance systems, and advanced concepts of the Transformational Communication Architecture.

In this chapter we survey the advantages of lasers over microwaves, emphasizing the higher bandwidth, narrower beamwidth, and smaller equipment size and weight. However, because of better weather penetration by microwave, a unique antenna design is presented: a combined microwave/laser antenna, which will simultaneously accommodate laser and microwave bands, thereby enabling, as a function of the operational environment, a shift from the higher band to the lower band and vice versa.

A summary of each chapter of the book is presented in Chapter 1. It is given in order to sketch out for both the student and the design engineer the essence of the entire book, as well as to help identify key features of the subject of laser space communications.

1.2 Advantages of Lasers over Microwaves

1.2.1 Narrow Beamwidth

The maximum narrowness of the laser beam is achieved with diffraction-limited optics, providing a beamwidth of

$$\theta = 2.24\lambda / D \qquad (1.1)$$

where

λ = wavelength of laser transmission
D = diameter of optical aperture of transmitting telescope

Clearly, comparing the laser beamwidth (e.g., λ = 1.0 micron with D = 10 cm yields 22.4 μrad) with that of a radio frequency (RF) signal (e.g., at X-band) would result in a much wider beamwidth. At 10 GHz (λ= 3 cm with D = 1.0m), the beamwidth will be 67.2 mrads.

Shown in Figure 1.1 is a beamwidth of 3°, based on the limited size of the antenna dish onboard the DSCS-2 satellite. The ground intercept on the Earth, when the antenna beam is focused directly down along the equator, would be a circle with a ~1880-km diameter, while the ground intercept of the laser, from a synchronous distance would also be circular, but with an 804-m diameter for λ = 1.0 μm and a 10-cm aperture. As shown, the small intercept permits transmission to a guarded and monitored area. Also by virtue of the smallness of the laser communication subsystem, several laser transceivers with their associated telescopes can be deployed on a single platform. Those transceivers can be placed on the lower and/or upper decks of the satellite platform.

As was indicated, the selected intercept areas illustrated in Figure 1.1 are circular, but clearly this is the case only when both the subsatellite point and the subbeam point are at the equator. Pointing the beam along the perpendicular off the equator will cause the circle to be stretched, so that it looks more like an ellipse that expands into large portions of the Earth and eventually falls off the Earth's surface. (Whatever the altitude of the satellite and its pointing angle of the antenna center, the resulting footprint on the Earth will be smaller than the RF case, provided both satellites have the same geometrical configuration).

As will be seen in later discussions, care is taken to ensure that the spacecraft supporting the laser communication subsystems is station kept. This is done as a first step in the design of accurate pointing and also, for certain applications, to ensure the maintenance of privacy to the intended callee's platform.

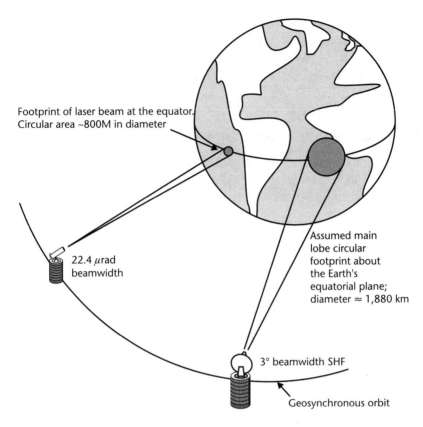

Footprint of laser beam at the equator. Circular area ~800M in diameter

22.4 μrad beamwidth

Assumed main lobe circular footprint about the Earth's equatorial plane; diameter ≈ 1,880 km

3° beamwidth SHF

Geosynchronous orbit

Figure 1.1 Privacy comparison between microwave versus laser footprints. Small intercept areas enable transmission to a private and monitored area. [Courtesy of the U.S. Government]

1.2.2 Large Directivity

Because of the very short wavelength (optical wavelength: 0.48–0.78 μm), very high directivity is attainable with small-size antennas. However, let us compare the laser wavelength with the microwave wavelength to demonstrate the advantage of the laser over the microwave in this regard,

$$\left(\lambda_{\text{laser}} / D_{\text{laser}}\right)^2 \div \left(\lambda_{\text{microwave}} / D_{\text{microwave}}\right)^2 = \theta_{\text{laser}}^2 / \theta_{\text{microwaves}}^2 \quad (1.2)$$

and the antenna directivity ratio may be expressed as

$$G_{\text{laser}} / G_{\text{microwave}} = 4\pi / \theta_{\text{laser}}^2 \div 4\pi / \theta_{\text{microwave}}^2 \quad (1.3)$$

As the beam is narrower for lasers, less power is required, and for diffraction-limited optics, the wavelength at optical wavelength is 10^3 to 10^4

smaller than microwave wavelengths. For example, at D = 10 cm, θ_{laser} = 10 μrad at λ = 0.78 μrad. However, at 10 μrad and λ = 0.78 μm, the antenna gain G = 109 dB, while at X-band and λ = 3 cm, the size of the antenna dish to achieve the same gain will have to be 10.16 · 10^5 cm (≅ 10 km). Clearly, this is an impractical size to be space deployed.

1.2.3   Higher Bandwidth of Lasers Versus Microwaves

A major advantage of lasers is their ability to transmit a much higher bandwidth signal than microwaves are able to achieve. For example, let us consider a 30-ps (30 · 10^{-12} sec) pulsewidth; its bandwidth should then be ≥ 39 GHz.

However, at a laser frequency of 3 · 10^{14} Hz (λ = 1 μm), the bandwidth-to-frequency ratio would be

$$\Delta v / v = 30 \cdot 10^9 / 3 \cdot 10^{14} = 10^{-4} \qquad (1.4)$$

Thus, 1,000 channels, each 30-GHz wide, could be accommodated at 10% of the optical carrier. That is,

$$(1000 \cdot 30 \text{ GHz}) / (3 \cdot 10^{14}) = 10\% \qquad (1.5)$$

By comparison with microwaves, 30 GHz will occupy the entire microwave band.

Although the commonly used rule of thumb when estimating a channel bandwidth is to take 10% of the carrier, there is the issue of the detector's bandwidth capacity. That is, the typical detector subsystem is able to process only a limited signal bandwidth. Therefore, it is necessary to determine the bandwidth limit to be used when assuming that a particular input signal bandwidth can be processed by the detector. Throughout, the channel bandwidth must be system consistent with the ability of the receiver to detect the entire bandwidth of the input signal. In other words, the necessary approach is working back from the detector bandwidth capacity to the design of the transmitted bandwidth.

1.2.4   Privacy Comparison Between Laser and Microwave

An indication of the privacy comparison is shown in Figure 1.2. The reference laser beam consists of a 1-arcsec beamwidth, and the microwave, 35-GHz signal, at 1/4° beamwidth. For the laser output beam, measuring it at a distance perpendicular to the beam's axis of 0.4 miles would result in roughly

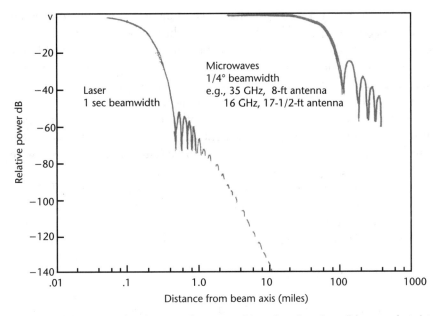

Figure 1.2 Privacy comparison between microwave and laser footprints. A small intercept footprint enables transmission to a private and monitored area.

40-dB down relative to the power from the output of the telescope. Moreover, at a distance of 10 miles astride the beam axis, the signal would be 140 dB down. However, the key point is that one must be quite close to the center of the beam (≤ 0.1 miles) to be able to listen without requiring a sensitive receiver.

For the case of the microwave signal of 35 GHz, the signal could easily be picked up at roughly less than 40 miles, and it would be down about 40 dB at ~100 miles. Clearly, this is a major loss of privacy.

A number of studies and measurements have been undertaken to provide design specifications for the laser-sensing signal receiver. It was concluded that one has to be physically close to the beam spot diameter on the ground to be able to detect the transmitted intelligence. As expected, it was also concluded that a much wider region of listening would be enabled by using the microwave signals.

1.3 Combined Laser and Microwave Communication

The benefit of having both microwave and light wave communications within a single antenna telescope is particularly evident onboard a ship, where poor weather conditions often exist, and superstructure space is limited.

As shown in Figure 1.3, we have a combination of two microwave frequencies and a number of laser lines. As seen, the parabolic reflector is located at one end of the telescope and the hyperbolic reflector at the other end. In the latter, the inner side has a metal surfaced mirror with slots to permit the f_1 signal to go through to the other end of the antenna. The f_2 will go through and be reflected by the hyperbolic section, whose face is totally reflective. Now with the different frequency carriers, we may configure a radar or a synthetic aperture radar (SAR), or simply a communication system. The latter is used when our transceiver is affected by weather that would result in significant loss were we to continue using one of the optical bands for communications.

Although communications in optical, near-wave infrared (NWIR), and long-wave infrared (LWIR) are being considered in this design, we can select lines within the three bands to accommodate a multiple number of different transceivers. All these bands will be employed in cloudless and other weather-free conditions. However, in foul weather we would have to shift down in bandwidth and transfer to the microwave bands, and a major laser advantage would be eliminated. Thus, as a general approach one would design the communication system to be able to shift down in frequency band as the weather conditions become poor.

Apart from the communication functions, as the laser and microwave bands would permit, passive sensors from the optical, NWIR, MWIR, LWIR, far infrared (FIR), and radiological bands can be used to provide considerable data from areas of interest.

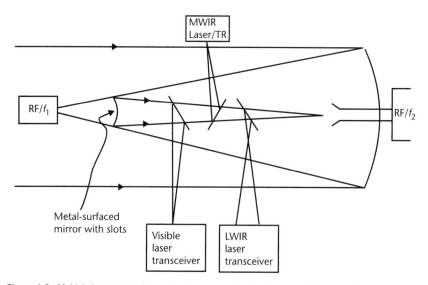

Figure 1.3 Multiple laser transceiver and microwave radar using the same telescope antenna.

1.4 Combined Signal-to-Noise for Microwaves in Uplink/Downlinks and Laser Crosslinks

For ground-based point-to-point communication on a worldwide basis, a combined microwave and laser communication link may be designed, involving a microwave uplink, a laser crosslink, and a microwave downlink.

As is indicated in Figure 1.4, we evaluate the total signal-to-noise ratio. We start from a transmitter on the ground, somewhere in the world, uplink to the first satellite, then crosslink from it to a second satellite via laser signal, and from that satellite to a second ground station, in another part of the Earth, via microwaves.

$$\mathrm{SNR}_{\mathrm{total}} = \frac{1}{1/(\mathrm{SNR})_{\mathrm{uplink}} + 1/(\mathrm{SNR})_{\mathrm{isl}} + 1/(\mathrm{SNR})_{\mathrm{downlink}}} \quad (1.6)$$

where

$\quad (\mathrm{SNR})_{\mathrm{uplink}} \quad = $ RF uplink S/N ratio
$\quad (\mathrm{SNR})_{\mathrm{isl}} \quad = $ intersatellite laser S/N ratio
$\quad (\mathrm{SNR})_{\mathrm{downlink}} \quad = $ RF downlink ratio

One may consider the following numerical example: For $(\mathrm{SNR})_{\mathrm{uplink}} = $ 25 dB and $(\mathrm{SNR})_{\mathrm{downlink}} = $ 17.8 dB, then $(\mathrm{SNR})_{\mathrm{isl}}$ of 22.9 dB is required, in order that the intersatellite link not degrade the $(\mathrm{SNR})_{\mathrm{total}}$ by more than 1 dB. An aspect of this example may assume that there are sensors onboard the first satellite, and therefore a large bandwidth of data is transmitted in the crosslink to the second satellite. Considerable data compression is undertaken to enable the RF downlink to handle the signal. The same data compression design approach may be taken when the combined laser/microwave

$$\mathrm{SNR}_{\mathrm{total}} = \frac{1}{\left(\dfrac{1}{\mathrm{SNR}}\right)_{\mathrm{uplink}} + \left(\dfrac{1}{\mathrm{SNR}}\right)_{\mathrm{isl}} + \left(\dfrac{1}{\mathrm{SNR}}\right)_{\mathrm{downlink}}}$$

- $\mathrm{SNR}_{\mathrm{total}} \quad = $ signal-to-noise ratio for the combined uplink, crosslink and downlink
- $\mathrm{SNR}_{\mathrm{uplink}} \quad = $ the microwave SNR uplink
- $\mathrm{SNR}_{\mathrm{isl}} \quad = $ intersatellite SNR laser crosslink
- $\mathrm{SNR}_{\mathrm{downlink}} \quad = $ microwave SNR downlink

Figure 1.4 Combined microwave uplink, laser crosslink, and microwave downlink.

antenna is used during poor weather. That is, when the atmospheric weather losses exceed a preset level, then the data compression process kicks in and transmission in the microwaves takes place.

To implement an extremely wide bandwidth laser communication system will require a system architecture that will include selection of particular lines from a selected laser, each of which is modulated by a separate information stream. The different beams are then combined in a optical antenna system, such as the Cassegrain telescope.

1.5 Review of the Chapters

1.5.1 The Essence of Chapter 2

Chapter 2 shows how to prepare the signal power budget and how to evaluate the effect of the physical vibration of the platforms. Varied details are provided, showing the components of the signal power budget and how to calculate the bit error rate (BER) for a selected modulation format. The pulse gated binary modulation (PGBM) is chosen in this text because it is easiest to implement circuitwise and is most commonly used in actual communication links. It is a form of on-off keying (OOK), but with better performance, as the receiver opens a gate when properly synchronized and coordinated with a transmitted pulse.

For a number of links described in Chapter 2 and Chapter 7, an analysis of the various components—power, laser wavelength, antenna (telescope) gain, range, external photon noise environment and internal receiver/detector noise, energy per photon, signal bandwidth, and margin—of the optical signal power budget will be presented, and expressions are derived for the required number of signal photons per bit and the corresponding number of noises per bit to achieve a given BER.

Unlike the typical RF case, the platforms' physical vibrations can produce a major increase in the bit error rate. That is because the narrowness of the beam (near diffraction limited, for example) going from the transmitter through the telescope will be displaced from reaching the center of the receiver-telescope. The displacement occurs because of the "jerkiness" of the platform when moving along its orbit, and is basically intermittent. Another cause involves the movements of certain pieces of equipment within the satellite that impact the outer envelope of the platform and are generally random in nature. While methods of ameliorating the effect of the vibrations are discussed in Chapter 3, a special vibration term indicating the pointing error of the laser beam of the transmitting satellite and its counterpart and the pointing error of the optical pattern of the telescope of the receiver satellite looking back at the transmitter need to be included in the signal power budget.

With the effect of the vibrations terms duly noted, calculations (shown in Chapter 2), will then be made for a standard crosslink between two synchronous satellites separated by 80,000 km and also for a low altitude satellite (LAS) articulating with a synchronous satellite and a range of ~40,000 km. Both the external photon noise, based on the satellites' spatial orientation and orbits (primarily sky noise plus solarshine, lunarshine, earthshine) and the internally generated noise due to the photon detector subsystem will be considered.

Because the BER is also a function of the modulation scheme, we demonstrate that fact by using PGBM and calculating the photon signal-to-noise ratio and associated BER for the direct detection (noncoherent) receiver (in Chapter 2). The results are compared with those for the BER for the coherent receiver systems: the heterodyne detection and homodyne detection systems. The problems with the coherent systems are that it is necessary to maintain phase coherency, and also one that needs to achieve practically zero distortion in the mixing process. Therefore, their use is appropriate in environments involving unique links.

As the mathematical simulation shows, with a high-efficiency mixing process, an advantage of as much as ~10 dB over the direct detection will be produced by the heterodyne receiver. But that has a number of caveats, apart from the short bounds on the phase distortion. For example, the designer will have to look at the environment in which the links are to operate as well the available optical power, the bandwidth requirements for the link, and the aliasing issues.

1.5.2 The Essence of Chapter 3

Chapter 3 discusses the major aspects of the acquisition tracking and pointing (ATP) architecture, including specific methods of taking out the effect of vibrations by using inertial instruments. The most common of these for purposes of illustration are the accelerometer subsystems, each of which would be connected to a corresponding coordinate axis of the satellite platform. Their output would be differenced with selected references, to servo-out the vibrations.

The example chosen for the ATP is the ubiquitous low-altitude satellite focused upward to articulate with the synchronous satellite. The basic steps of ATP in this process include the use of a beacon turned on by the synchronous (relay) satellite, with the LAS responding. When the upper altitude platform detects closing of the link, the high data rate begins to flow from the LAS to the synchronous (relay) satellite. Clearly, because the LAS is moving along its orbit faster than the relay satellite, you need to have a point-ahead angle (PAA) beacon signal to close in on the LAS. Actually, the orbital position of the LAS may be obtained from the global positioning system (GPS) satellite, and this information is transmitted to the relay satellite.

A number of diagrams are presented covering these discussions, together with additional material associated with the broadbeam and scanned-beam methods of acquiring the platforms and the acquisition time sequence. A number of design equations are derived and the latest state-of-the-art circuitry is indicated.

1.5.3   Overview of Chapters 4, 5, and 6

In Chapters 4, 5, and 6 we consider selected probabilistic models of the atmospheric turbulence for both downlink and uplink cases. Also included are the modified signal power budgets and examples.

In Chapter 6, Beer's Law is used to determine the laser signal losses due to the effect of molecular and aerosol absorption and scattering. We also evaluate the effect of weather, such as rain, fog, clouds, and snow, on the signal transmission. Tables of data sources from LOWTRAN and MODRAN software systems, which could be used in the evaluation of the index of refraction structure constant and assorted signal loss components, are indicated. In this chapter, the transmission of the laser beam signal over terrestrial links at distances of 30 km is evaluated. An example is also given of longer distances (148 km), which depend on having tall towers at either end of the above-mentioned ground link. An observable line of sight between the high towers will clearly increase the range between the transmitter and receiver. While the terrestrial range is limited by the curvature of the earth, the taller the towers upon which the transmitter and receiver telescopes are placed, the longer will be the line-of-sight distance (between the optical transmitter and receiver), at which widespread communication with a low BER can be sustained.

Chapters 4, 5, and 6 may be considered as a single group because the laser communication beam interacts with the atmosphere and requires special turbulence analytics, as well as the physics of absorption and scattering, in their technical assessment.

1.5.3.1   The Essence of Chapter 4

In Chapter 4, we evaluate the downlink from a satellite to an optical ground station. Equations describing signal losses due to the atmospheric turbulence and weather conditions are evaluated. The BER of the downlink signal through the atmosphere for OOK modulation is calculated.

To mitigate the effect of atmospheric turbulence, an adaptive optics (AO) subsystem in the optical telescope, together with a reference laser beam, may be employed. One can measure the beam's distortion when going through the atmosphere and compensate for the distortion by making adjustments in the deformable mirror of the AO. The output laser signal that has been distorted by the deformable mirror will, when interacting with the distortion in the atmosphere, tend to cancel a lot of the effect of distortion that has been introduced into the signal path by the downlink atmospheric turbulence.

Although the atmospheric turbulence, when combined with the laser beam jitter, can substantially increase the BER when it is combined with the "zeroing out" of the jitter by the method described in Chapter 3, together with implementing the AO technology, significant reduction in the BER can be achieved. This performance is described by the calculations shown in Chapter 4.

The basic analysis of penetration of a laser beam through the atmosphere using complex relationships was first introduced by Tatarsky. But these calculations were reduced by Fried and, principally, by Yura to more practical engineering approximations with simpler expressions that are more readily calculated. Those relationships with experimental supportive data are plotted, for ease of use, with the communication systems' architectures presented in Chapters 4 and 5.

1.5.3.2   The Essence of Chapter 5

As explained in Chapter 5, as a laser beam goes down from a satellite platform to a ground station, the beam spreads as a product of the beamwidth at the exit of the optics and the distance to the ground station. Thus, down to the last 30 km from the ground, the beam is spread geometrically until encountering the measurable atmosphere, which will cause it to spread a bit more. However, when going up from the ground station to the satellite, the beam is spread immediately to a broader width because of the near-in atmospheric turbulence. However, through the use of an adaptive optics subsystem (AOS) in the telescope, with an appropriate input reference, the effect of atmospheric turbulence can be abetted. In addition to using AOS to correct phase distortion, aperture averaging can be employed to reduce turbulence-induced scintillations.

Expressions are developed to indicate the measure of the signal coupling efficiency (SCE) as a function of the lateral coherence distance, which in turn is a function of the zenith angle, the atmospheric index constant, and other parameters. Thus, when going through the atmosphere, the expression of signal power budget, as derived in Chapter 2, needs to be multiplied by the factor of signal coupling efficiency in order to get a measure of the uplink loss.

To return to the design of the adaptive optical system, in a number of instances the required reference laser signal may be gotten by exciting the sodium layer of the atmosphere which is located at an altitude of ~90 km. This excitation source is derived from a ground-based, pulsed dye laser (λ = 0.589 μms) or by having a separate reference laser located on the satellite and aimed at the optical ground station. A reference laser may also be based on a separate satellite platform. Such a satellite may also support a relay mirror, which could reflect the laterally aimed laser data signal from a sensor satellite to the ground station or some other platform. Finally, the reference may be initiated from a ground-based laser (GBL) going up to the relay satellite, which also has a retroreflector mirror for retransmission to the ground-based station's adaptive optics subsystem. More details about the reflective mirror structures are discussed in Chapter 8.

Chapter 5 concludes with a description and analysis of what the signal loss would be if the adaptive optical system were not in place. As shown, losses on the order of 18–20 dB are likely, even if the tower supporting the telescope is a kilometer above sea level and the height of the tower is 10m.

1.5.3.3   The Essence of Chapter 6

Chapter 6 begins with the design of a laser link between two towers that are separated by about 30 km from one another and are 40m in height. This is followed by another overhead link that is 148 km long. These line-of-sight, atmospheric links have so far been developed for experimental purposes, with measurements still being undertaken in the program. In terms of link performance during inclement weather, such as dark clouds, fog, and rain, significant losses are induced on the laser beam. However, an approach has been developed that tends to get around these weather issues.

Titled "Terrestrial Laser Communication Links and Weather Issues," Chapter 6 presents experimentally obtained data losses due to clouds, fog, rain, and snow. We also present a weather avoidance system (WAS), which demonstrates a way of bypassing inclement weather by interconnecting our optical ground stations with underground and overhead fiber-optic cables. Thus, by getting near-real-time data from weather satellites and other National Oceanic and Atmospheric Administration (NOAA) sources, we can point the space-based telescope antenna to a ground station that is located in a dry and clear environment. This downlink signal can then be transferred via the fiber cable to a desired station, which may located in a harsh weather environment. This station may be a command center, for example, which would otherwise be unable to be in laser communication with some vital space assets.

1.5.4   The Essence of Chapter 7

An advanced Internet system and associated protocols have been developed and are presented in Chapter 7. It is named the Fifth Generation Internet (5-GENIN) System. It includes a synchronous backbone that is composed of three synchronous satellites that are separated at 120° about the center of the Earth. In this fashion, all uplink signals go, directly or by way of relay nodes, to the synchronous backbone from ground-based stations, airborne platforms (fixed and rotary winged), and robotic mobile ground elements such as the Miniaturized Unmanned Ground-Based Mobile (MUGM) System. The signals (generated by the callers), depending on their addressees, will be transferred to different circuits of the synchronous backbone for specific downlink paths, to the intended callees.

Uplink and downlink from submerged vessels would also be part of the 5-GENIN communication system configuration. Moreover, future interplanetary nodes established by the National Aeronautics and Space Administration (NASA), and some NOAA weather satellites' interconnectivity with ground stations can also be part of this advanced network.

In a related portion of Chapter 7, the WAS is discussed. Use is made of the available ground-based optical fiber nets (which may be located in overhead and underground cables and also in submarine cables that interconnect the optical ground stations). As we will see, the system concept is applicable to all types of worldwide communications. In this book, WAS is woven into the 5-GENIN laser system architecture.

1.5.5   The Essence of Chapter 8

In Chapter 8 we describe two passive spatial structures that are capable of reflecting a transmitter laser signal to a particular receiver on the ground. The technical feasibility of enabling a mirror in space to reflect an uplink or a crosslink signal to a ground station is based on results garnered from the Remote Mirror Experiment (RME) System measurement programs which were made in the late 1980s. The results were published in 1991. Apart from the relaying aspects of the spatial mirror, the reflector that is used in its retroreflective modality can help to provide a reference for a ground-based AOS. In other applications, a mirror with a highly reflective surface that is also heat resistant can be effective in reflecting a high-energy laser beam toward a selected target. This approach may also be used in a laser radar system and, when used with a blue-green laser system, can be helpful in penetrating sea water, in search of underwater vessels, for example.

Apart from the articulating mirror, the other reflective structure considered in Chapter 8 is the Optical Westford System. The approach used in evaluating this structure's reflectance performance is by calculating the signal power budget for the systems and comparing the result with a standard or "reference downlink signal" of 10^9 bits per second, an associated BER of 10^{-7}, and a modulation format of pulse gated binary modulation.

It is significant that the RME program demonstrated that a mirror controlled in space can be an effective tool in directing a laser communication beam to a particular station on earth, in the atmosphere, or in space. Moreover, it serves as an extrapolator for directing laser energy to a selected target in space or the atmosphere. Other passive reflective structures and their applications may also be inspired by the RME.

1.5.6   The Essence of Chapter 9

The last chapter of the book, Chapter 9, highlights special applications of laser beam communications. They include laser communication through sea water, interplanetary laser communications, and laser communication subsystems using microsatellites.

First we describe the use of a blue-green laser to penetrate sea water for communications to underwater vessels and returning communications from the submarines to the satellite or an atmospheric platform. We also describe a space-based antisubmarine warfare (ASW) system concept, called the space-based detectability and identification of submersibles system (SBDIS).

Although the Lunar and Martian laser communication programs have been canceled due to funding issues, because calculations have indicated their links' feasibility, the essence of these links are described. They include a laser link from a Lunar station to an Earth station (or a synchronous satellite) and from a Martian station (or a low-altitude satellite orbiting Mars) to an Earth station (or a synchronous satellite orbiting the Earth). However, at this time period microwaves, using the X-band and Ka-band as backup, are used to downlink to Earth.

Finally, we discuss two subsystems in the microtechnology domain, in which very small-sized laser transceivers are considered. One is known as the Steered Agile Laser Transceiver (SALT) and the other is the retroreflective communication system (RRCS), which uses the multiple quantum well modulator. Both subsystems are currently in laboratory development. However, they may be deployed within a few years in a variety of satellites, including nanosatellites, and in unmanned airborne platforms.

Finally, a brief overview is given of the Transformational Satellite (TSAT) System, wherein a number of wideband links are interconnected to achieve time-orderly communications from varied sources to selected command centers or in the caller- or callee-type communications linkages from all system grid nodes. This represents part of the early phase of the TSAT effort within the context of the transformation communication architecture (TCA).

2

The Signal Power Budget for Intersatellite Links and Potential Mars-to-Earth Links

2.1 Introduction

This chapter discusses the generic laser signal power budget (SPB) that is required between any two satellites orbiting in the nonatmospheric environment, in order to achieve a given BER for the selected modulation format. Also outlined is the potential Mars-to-Earth laser link.

In the derivation of the laser SPB for the communication links, it is assumed that the optical beams are firmly locked in place from the transmitter satellite to the receiver satellite. In practice, this could not be the case. In fact, considerable design effort is necessary to attain the proper point-ahead angle between the satellites, as well as in carrying out the necessary acquisition, tracking, and pointing (ATP) processes. This would allow the communication beam to "lock" on to the receiver satellite for the specified period of communication. The vital subject of ATP is discussed in Chapter 3. There we demonstrate how the basic SPB is modified by adding factors that indicate the changes in position of the transmitted optical beam and the receiver optical pattern, due to natural and induced vibrations of the satellite platforms. Both systemic and random vibrations are considered, and as shown in Chapter 3, both can be taken out by means of dedicated servo loops. The details of the vibration issues and their effects on BER are discussed in Sections 2.6 and 2.7.

For communications in the nonatmospheric medium between any two satellites, the laser SPB is similar to the signal power budget when RF is employed. However, in the laser communication case, we convert from the signal in watts per cycle (or bit) to a signal in number of photons per cycle (or bit), and the noise power in watts per cycle is converted to the number of noise photons per cycle. That is, we obtain a ratio of signal photons to noise photons, per each information cycle (or bit).

We use the expression hv for the energy per photon in units of joules (watt-seconds) per photon, where h (Planck's constant) equals $6.625 \cdot 10^{-34}$ in units of watt-second (joule) per photon per Hz, and frequency v of the laser light is measured in Hz. Thus, n = P/hvf. That is, when dividing the received optical power (in watts) by hv and by the signal data rate f (in bits per second), the number of photons, n per bit, is obtained.

The discussion of the noise photons will be given in a later section of this chapter. However, the point to be made here is that there is a simple method of evaluating the number of collected photons at the receiver satellite, using the optical power emanating from the transmitter satellite. It involves primarily the optics onboard the two satellites' platforms, the laser power output, and the photo-detector subsystem at the receiver.

In the receiver design, a photoelectric device such as an Avalanche Photo Diode (APD), P-Intrinsic (PIN) photodiodes, a photomultiplier tube, or another kind of photocell system is used to produce the desired signal in photoelectrons. An electronic filter is then utilized to optimally separate the signal photoelectrons from the noise photoelectrons.

2.2   Signal Power Budget Calculation [1]

As seen in Figure 2.1, the received power, P_R, is equal to the product of P_T, the transmitted power, multiplied by the ratio of the solid angle of the receiver, subtended at the transmitter aperture, to the solid angle of the transmitter into which P_T is fed. That is,

$$P_R = P_T \Omega_R / \Omega_T \qquad (2.1)$$

Ω_R = solid angle of the receiver, subtended at the center of the transmitter aperture, $\pi(a_R)^2 / R^2$,

where

  R   = distance between the transmitter and receiver optics
  a_R = radius of receiver aperture
  a_T = radius of transmitter aperture
  Ω_T = solid angle of the transmitter into which P_T is fed, = $\lambda^2 / \pi(a_t)^2$

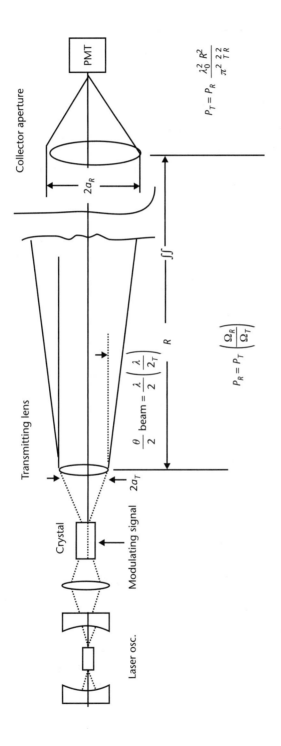

$$\frac{\theta}{2} \text{ beam} = \frac{\lambda}{2}\left(\frac{\lambda}{2_T}\right)$$

$$P_R = P_T\left(\frac{\Omega_R}{\Omega_T}\right)$$

$$P_T = P_R \frac{\lambda_0^2}{\pi^2} \frac{R^2}{a_T^2 a_R^2}$$

P_R = Received power
P_T = Transmitted power
Ω_T = Solid angle of optical transmitter into which P_T is fed, $\dfrac{\lambda_0^2}{\pi a_T^2}$

Ω_R = Solid angle of receiver, subtended at the center of the transmitter aperture, $\dfrac{\pi a_R^2}{R^2}$

Figure 2.1 Calculations of transmitter power.

By substitution, we have

$$P_R = P_T \pi^2 \left(a_R\right)^2 \left(a_T\right)^2 / \left(R^2\right)\left(\lambda^2\right)$$ (2.2)

Also, since P_R is the received optical power, it is equivalent, in terms of the number of received photons per bit, to

$$P_R = n(h\nu)f \text{ or } n = P_R/(h\nu)f$$ (2.3)

To obtain the number of signal photoelectrons: n´, we use the photoelectron detector with a quantum efficiency of Q , to obtain

$$n' = Q_n = Q P_R/(h\nu)f$$ (2.4)

where Q, the quantum efficiency of the photoelectron detector, is the ratio of the output photoelectrons per input photons.

To continue in the development of the SPB, we add the factor F, to represent the combined efficiencies of the transmitter and receiver subsystems, that is, $F = L_T \cdot L_R$. In addition, M = Margin is also included as a factor. It should be noted that M is often called the safety factor (SF) of the design of the communication link.

To detail the expression for n´, (2.2) and (2.4) are combined, giving

$$n' = P_T \left(D_R\right)^2 (FQ)/M\left(R^2\right)\left(\theta^2\right)(h\nu)f$$ (2.5)

where θ = optical beamwidth of the transmitter.

The other symbols shown in the relationships of the optical SPB are presented in Figure 2.2.

2.2.1   Numerical Example [1, 2]

Let us assume that the laser employed as the transmitter in the intersatellite link is the Nd:YAG, which produces light at 1.064 µm, and that the required photoelectron per bit is 40 (or 16 dB). This, as will be seen later in the chapter, is the required n´ (photoelectrons) to achieve a 10^{-8} error rate for pulse gated binary modulation (PGBM) with an extinction ratio of 20 dB. However, with the detector's quantum efficiency of 30% (5.2 dB), it will yield 133 (or 21.2 dB) as the required number of photons per bit.

With the energy per photon (h = $6.625 \cdot 10^{-34}$ joules per Hz per photon, at λ = 1.064 µm and a frequency of $3 \cdot 10^8$ / $1.064 \cdot 10^{-6}$ yields ~ $3 \cdot 10^{14}$ Hz), hν = 187.2 dB joules per photon. Further, since the number of required photons per bit = 133 (21.2 dB) to achieve the required BER (as

$$P_R = \frac{n(h\nu)f}{Q}; \quad n = \frac{P_R}{(h\nu)f}Q$$

$$P_R = \frac{P_T d_R^2 F_L}{MR^2\theta_T^2} \quad P_T = P_R \frac{\lambda^2 R^2}{\pi^2 \alpha_T^2 u_R^2}$$

$$n' = \frac{P_T d_R^2 F_L Q}{MR^2\theta_T^2 (h\nu)f}$$

n = required number of signal photo-electrons per bit; it is a function of modulation scheme and background pe for a given error rate

$h\nu$ = energy per photon

Q = quantum efficiency

f = data rate in bits per second

P_R = received power

P_T = transmit power

F_L = combined efficiencies of transmitter and receiver

θ_t = optical beamwidth (transmitter optics)

R = range between transmitter and receiver

M = margin

Figure 2.2 Relationships of optical signal power budget.

already indicated), the total number of joules per bit is $h\nu$ = −187.2 dB (joules per photons) + 21.2 dB photons per bit = −166 dB (joules per bit). Finally, for a signal data rate of 10^9 bits per second, the value of the joules per bit is multiplied by the data rate, giving −166 dB (watt-sec per bit) + 90 dB (bits per second) = −76 dBw, which is equivalent to $2.5 \cdot 10^{-8}$ watts per bit.

In Figure 2.3 and Figure 2.4 the same power budget is calculated for the GaAlAs semiconductor laser, but with the latter's data rate of 12.6 megabits per sec. The GaAlAs is assumed to be the beacon laser source in this example, and its wavelength output is 0.780 μm.

		Nd:YAG	GaAlAs diode laser
• Required photelectrons/BIT	dB	16 (40)	16 (40)
• Detector quantum efficiency	dB	5.2 (30%)	5.7 (0.27)
• Required photons per BIT	dB	21.2 (133)	21.7 (148)
• $h\nu$ = watt-sec/photon	dB	−184.2	−185
	dBJ	−163.	−163.3
• Joules per BIT = 21.2 − 184.2	dB	$90.(10)^9$	$70.9(12.6 \cdot 10^6)$
• Bit rate (assume 1 GBPS);		− 73 dBW	−92.4 dBw
10^9 bit/sec		$(5 \cdot 10^{-8}$ watts)	$(5.75 \cdot 10^{-10}$ watts)

- Watts required at detector, P_{RD}
- 10^{-7} BER, 1 Background pe/decision period, PGBM; extinction ratio −20 dB
- h = $6.625 \cdot 10^{-34}$ watt-sec/Hz
- ν = c/λ = $3 \cdot 10^8/0.53 \cdot 10^{-6}$ = $6 \cdot 10^{14}$ Hz

Figure 2.3 Link power budget calculations of satellite-to-satellite link measuring 45,000 nmi (81,000 km).

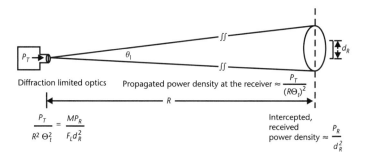

P_T

Diffraction limited optics   Propagated power density at the receiver $\approx \dfrac{P_T}{(R\Theta_t)^2}$

θ_1

d_R

R

$$\frac{P_T}{R^2\,\Theta_t^2} = \frac{MP_R}{F_L d_R^2}$$

Intercepted, received power density $\approx \dfrac{P_R}{d_R^2}$

P_R = Power required at the detector (5 × 10^{-8} watts)	−73.0 dBw (.Nd: YAG)	−92.4 dBw (diode laser)
R = Intersatellite distance 8.33 × 10^7 meters; R^2	158.4	158.4
Θ_T = Transmitter's diffraction limited optics; Θ_t^2	−106.0 (θ = 5 × 10^{-6} radians)	−106.0
d_R = Collector optics, 0.5 meters; d_T^2	−6.0	−6.0
F_L = System efficiency, 25%	−6.0	−6.0
M = Margin	5.0	5.0
P_T = The transmitter optical power	−3.6 dB (0.44 watt)	−23.dBw(0.5 mW)

Figure 2.4 Link power budget, 45,000 nmi (81,000 km) satellite-to-satellite link (data and beacon).

The diode laser is likely to be used because of long-term performance reliability compared to other lasers, particularly the HeNe laser, which was used in earlier designs. The GaAlAs wavelength is of the order of 6% of the wavelength of the HeNe laser. Thus, the calculations made for one of the wavelengths will be acceptable for the other. With today's technology, however, the fiber laser is more likely to be used. It generates 1.550 μm, which is a wavelength carrier commonly utilized in fiber links constituting the Earth's cable transmission.

2.3 Summary of the Power Budget Expression

A more conveniently expressed power budget relation that explicitly presents the basic components of the communication link between the two satellite platforms is given in (2.6) [1]:

$$n' = P_T L_T G_T G_R L_R Q \left(L_{P-T}\right)\left(L_{P-R}\right)/L_S\left(h\nu\right)f \tag{2.6}$$

where

n' = number of photoelectrons per bit
P_T = laser optical power output from transmitter
L_T = total signal losses in the transmitter system
G_T = gain of the transmitter antenna = $(\pi D_T/\lambda)^2$
D_T = diameter of the transmitter aperture

G_R = gain of the receiver optical antenna = $(\pi D_R/\lambda)^2$
D_R = diameter of the receiver antenna
L_R = total signal losses in the receiver system
f = frequency of the data stream in bits per second
L_S = free space loss = $(4\pi R)^2/\lambda^2$
R = range between transmitter and receiver
h = Planck's constant
v = frequency of the laser light
hv = energy per photon
Q = quantum efficiency
L_{P-T} = pointing loss of the transmitting beam
L_{P-R} = pointing loss of the receiver's optical antenna beam

The expressions for L_{P-T} and L_{P-R} will be derived in the last section of this chapter, where we will consider the effect of the platforms' physical vibrations on the pointing loss of the transmitted signal beam and also on the received beacon beam. These vibrations cause the transmitted beam to move away from the center of the receiver's telescope antenna. That is, the vibrations' amplitude and frequency that are superimposed on the beam cause it to move away from the center of the collector telescope antenna on the receiving satellite, thereby resulting in an increase in the BER of the communication link. Clearly, the effects of vibrations are particularly severe when the distances between the two articulating satellites are relatively small and the beam is very narrow, for example, of diffraction limit quality. In the case where the articulating platforms are a relatively large distance from one another, the transmitted beam is spread over a large diameter, hence the superimposition of the jitter may not be effective enough to move the beam away from the receiver aperture.

2.4 Evaluation of BER as a Function of Photoelectrons per Bit and Modulation Scheme

An example of the evaluation of BER as a function of the number of photoelectrons per bit for the very useful modulation scheme of PGBM is presented in Figure 2.5.

PGBM is a logical extension of the utility of the power budget covered in Sections 2.1 and 2.2. Modulation schemes such as pulse polarization binary modulation (PPBM), and pulse position modulation (PPM), and on-off keying (OOK) modulation are discussed next, along with a brief presentation on the direct and heterodyne receivers.

Shown in Figure 2.6 is an example of the waveforms for data word 11010, using PGBM. As indicated, this is a one-bit-per-pulse stream that is

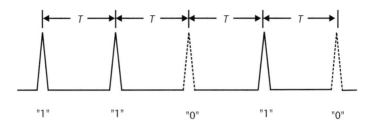

- One bit/pulse modulation format
- Requires high speed modulator
- Ideally suited for mode-locked laser transmitter
- Compatible with pulse gated receiver
- Achieves ideal noise discrimination

Figure 2.5 Pulse-gated binary modulation.

ideally suited for the mode-locked operation of the laser, enabling it to handle a high-speed modulator. By being compatible with a pulse-gated receiver, its discrimination against noise is high.

2.4.1   Error Analysis for PGBM [2]

The following discussion leads to an expression for the probability of error, for a given number of signal photoelectrons, background (noise) photoelectrons,

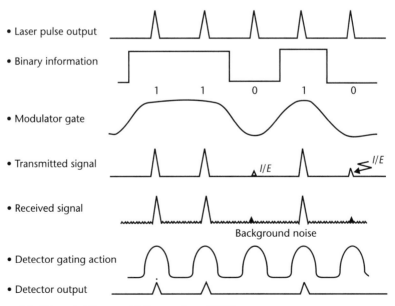

Figure 2.6 Pulse-gated binary modulation waveform: 11010.

and extinction ratio. (The extinction ratio, E, is the ratio of the number of signal photoelectrons received when a pulse is desired to the number of photoelectrons received when no pulse is desired.)

The probability of error may be expressed as

$$P_E = pP_{nd} + (1-p)P_{fd} \qquad (2.7)$$

where

p	=	probability of transmitting a pulse
(1 – p)	=	probability of not transmitting a pulse
P_{nd}	=	probability of no detection of transmitted pulse at the receiver
P_{fd}	=	probability of false detection (i.e., probability of detection of a pulse when the pulse was not transmitted)

Since the signal and noise are Poisson distributed [3], the following equation describes the probability of no detection at the receiver when a pulse was transmitted

$$P_{fd} = \Sigma_{k=T}^{\infty} \left(\overline{m}_B\right)^k E^{-\overline{m}_B} / k! \qquad (2.8)$$

where

T	=	optimum threshold (maximum likelihood ratio detection)
$\overline{m}_S$	=	mean number of signal photoelectrons per decision period
$\overline{m}_B$	=	mean number of background (noise) photoelectrons per decision period

and the probability of no detection of pulse in a decision period is

$$P_{nd} = 1 - \Sigma_{k=T}^{\infty} \left(\overline{m}_S + \overline{m}_B\right)^k E^{-(\overline{m}_S + \overline{m}_B)} / k! \qquad (2.9)$$

The following equation defines P'_{fd}, the probability of false detection for E, a finite extinction ratio. That is, the receiver for a given extinction ratio determines that a pulse is received when no pulse was transmitted:

$$P'_{fd} = \Sigma_{k=T}^{\infty} \left(\overline{m}_S / E + \overline{m}_B\right)^k E^{-(\overline{m}_S / E + \overline{m}_B)} / k! \qquad (2.10)$$

Plots of the error rates for different values of background photoelectrons, an extinction ratio of 100, and an average number of signal photoelectrons from 0 to 100 are presented in Figure 2.7, with the resulting BER, P_E, ranging from 0.02 to 10^{-8}.

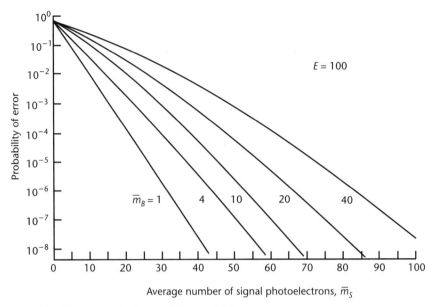

Figure 2.7 Bit error rate of pulse-gated binary modulation.

2.4.2 Example of Noise Background Calculations

Background radiance interferes with the desired signal photons. The Moon, stars, sky, and also earthshine produce interference affecting the signal photons. However, these are typically about 40 dB below the level of solar radiance. Specifically, the earthshine is 0.013 times the solar irradiance.

It has been determined that the solar background would produce 15 photons at the satellite receiver at 1.06 μm (Nd:YAG), with a detector quantum efficiency of ~40%. However, at 0.80 μm and Q = 22%, 7,000 photoelectrons would be detected, and at green light (0.53 microns), 33 photoelectrons would be detected at a quantum efficiency of Q = 60–70 %.

2.4.3 Detailed Background Calculations

The background power due to the Sun is obtained directly from the relation

$$P_B = \text{(Solar Irradiance)} \cdot \text{(Area of Receiver Aperture)} \cdot \qquad (2.11)$$
$$\text{(FilterBW)} \cdot \text{(Solar-FOV)}$$

Inputting the following five numerical values into Equation (2.11) will yield the background power due to the solar irradiance:

1. The solar irradiance is $7 \cdot 10^{-2}$ watts/cm^2/μm at 1.06 μm.
2. The receiver aperture is $\pi/4 \; d^2 = \pi/4 \; (50 \text{ cm})^2$.
3. The filter bandwidth is 1A: $10^{-4} \mu$m $= 10^{-10}$ m.
4. The solar field of view (FOV) is (diameter of the Sun) / $93 \cdot 10^6$ miles $= 9 \cdot 10^{-3}$ rad. This is equivalent to $(\pi/4) (9{,}000)^2 \mu$sr.

Thus, $P_B = 1.08 \cdot 10^{-8}$W. For a Gb/sec data stream, one bit period of 10^{-9} sec will produce $1.08 \cdot 10^{-17}$ J. For a gate period of 0.25 ns, the amount of energy per bit will be $\sim 0.27 \cdot 10^{-17}$ J.

Finally, as the energy per photons at 1.06 μm is $1.87 \cdot 10^{-19}$ joules, the number of solar background photons will be $1.7 \cdot 10^{-17}$ divided by $1.87 \cdot 10^{-19}$, which yields ~ 15 photons with the 0.25-ns gate. A summary of the background calculations just shown is given in Figure 2.8.

2.5 Direct Detection Versus Heterodyne Detection

As is common in all communication systems, it is desirable to calculate the signal-to-noise ratio, which is one of the important measures of the performance of the receiver and can be used for comparison with different receiver designs. The direct detection receiver is simple in its design, has fewer components than the coherent receiver, and does not depend on the phase of the signal. Its essence is to collect photons and identify whenever more then several photons are received per bit, which would indicate a 1. When fewer photons are collected, a 0 is indicated. Noise should be minimized, so that one can easily differentiate the noise level of 0 from that of a 1.

In the heterodyne receiver or coherent receiver design, there is an advantage of reducing the accumulated noise by means of having the local

P_B = (Solar radiance)(Area of Receiver Aperture)(filter)(Field of View Factor)
  = $(7 \times 10^{-2})(2025)(10)(20/9000)2 \simeq 1.08 \times 10^{-8}$ watts
The sun subtends a FOV of about 9 milliradian at the earth:

- Solar Radiance: 7×10^{-2} watts/cm^2/micron, at 1.06 micron
- Area of Receiver Aperature = $\pi/4(d^2) = \pi/4(50 \text{ cm})^2$
- Filter = 1 Angstrom = 10^{-4} micron = 10^{-10} meter
- Receiver FOV = 4 sec = 20 microradians (planar); $\pi/4(20)^2 \mu$ster
- Solar FOV = $D_{SUN}/93 \times 10^6$ miles $\simeq 9$ milliradians; $\pi/4(9000)^2 \mu$ster

Figure 2.8 Background calculations.

oscillator power exceed the noise power and the signal power. However, there is a need for the phase of the local oscillator output to be in phase with the input signal coming out of the filter into the mixer. Thus, any phase distortion in the optical signal arriving at the receiver will introduce a higher bit error rate, when compared to the direct detection receiver.

2.5.1 Signal-to-Noise Ratio for the Direct Detection Receiver

Starting with P_C, the optical power collected by the receiver, the photon rate is obtained by dividing P_C by $h\nu$, the energy per photon. That ratio will indicate the collected number of photons per sec:

$$P_C/h\nu \qquad (2.12)$$

(Dimensionally, the expression is watts/joules/photons/Hz · Hz, or photons per second.) And as noted earlier, Q, the quantum efficiency of the detector surface is equal to the ratio

$$Q = \text{output photoelectron rate/input photon rate} \qquad (2.13)$$

The signal current may then be represented by

$$i = q(\text{photoelectron rate}) = qQ\ (\text{photon rate}) = qQP_C/h\nu \qquad (2.14)$$

The signal power, P_S, can be expressed, after current multiplication and being fed into the load resistor R_L, as

$$P_S = \{G\}\,R_L = \{GqQP_C/h\nu\}\,R_L \qquad (2.15)$$

where G = photoelectric current gain.

The noise components generated in the receiver will be primarily due to the Schottky shot noise and the thermal noise. The shot noise power is due to i_{SS}, the signal current, i_{BS}, the background current; and the dark current, i_D. Thus, the total squared noise current is equal to the squared shot noise currents and the squared thermal noise current:

$$i_{\text{total}}^2 = i_{SS}^2 + i_{BS}^2 + i_D^2 + i_{nT}^2 \qquad (2.16)$$

where

$$i_{SS}^2 = \left(2qQ\{G\}P_C B_n\right)/h\nu$$
$$i_{BS}^2 = \left(2qQ\{G\}P_B B_n\right)/h\nu$$

$$i_D^2 = 2qi_D B_n$$

$$i_{nT}^2 = N_o B_n$$

B_n = noise bandwidth

N_o = electronic thermal noise spectral density in (watts) /Hz

Summing the previous four equations will yield for the total noise power, i_{total}^2:

$$i_{total}^2 = 2q\{G\}QB_nR[P_C + P_B] + 2qiB_nR + 4kTB_n \qquad (2.17)$$

Dividing (2.15), the signal power, by (2.17), the total noise power, will yield the signal-to-noise ratio (S/N):

$$S/N = \left(GqQP_C / hvR_L \Bigg/ \begin{cases} 2qGQB_nR_L(P_C + P_B) \\ +2qi_D B_n R_L + 4kTB \end{cases}\right) \qquad (2.18)$$

When P_C, the signal optical power, is much larger than the noise components, that is,

$$P_C \gg P_B, \; 2qiB_nR, \; 4kTB_n$$

then (2.18) reduces to the quantum limit performance of the receiver:

$$S/N = QP_C / 2B_n(hv) \qquad (2.19)$$

The generic block diagram of the direct detection receiver is shown in Figure 2.9. As seen, there is simplicity of architecture based on the fact that the

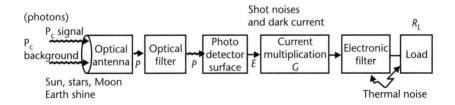

$$\frac{S}{N} = \frac{\left(\dfrac{qQG^2}{hv}\right) R_L P_c^2}{2qB_oG^2\left\{\dfrac{Qq}{hv}[P_c + P_B] + I_D\right\} R_L + 4kTB_o} = \frac{QP_c}{2B_ohv_c} \quad for \; P_c > P_B, P_D, 4kTB_o$$

Figure 2.9 Generic block diagram of direct detection laser receiver.

receiver is designed to collect photons without concern as to phase. As shown in the next section, the coherent receiver design requires many more components and more complex circuitry, with special attention paid to phase and polarization matching between the input optical signal and the optical local oscillator. But this more complex receiver has the advantage of achieving a higher S/N, implying a lower BER, as well as being able to detect phase and frequency.

2.5.2 Signal-to-Noise Ratio of the Heterodyne (Coherent) Receiver

In this section, the expression for the S/N of the coherent optical receiver is presented, having the implicit requirement that the input optical signal and that of the output of the optical local oscillator are nearly in phase, or within the quantity represented by the wavelength of the signal divided by the aperture diameter. There is a loss due to this phase difference and also a loss due to the polarization mismatch between the input signal and the local oscillator. There is also a loss due to the mixing process: it is expressed in the S/N equation as the mixing efficiency.

$$S/N = \left\{ P_M L_P M_E \left(GqQ/h\nu \right)^2 P_{LO} P_C R_L \right\} \div \qquad (2.20)$$

$$\left[qG^2 B_n \left\{ qQ/h\nu \left(P_C + P_{LO} + P_B \right) \right\} R_L / 2kTB_n \right]$$

where

P_M = phase match factor
L_P = loss due to polarization mismatch
M_E = mixing efficiency

When the local oscillator power is larger than the input signal power, the background power, the dark current, and the thermal noise, that is

$$P_{LO} > P_C, P_B, P_D, 2kTB$$

with the assumption that P_M, L_P, and M_E are all 100%, then equation (2.20) is reduced to the quantum limited performance:

$$S/N = QP_C / \left(B_n \right) \left(h\nu \right) \qquad (2.21)$$

Comparing this equation with that of the S/N for the direct detection receiver, it is seen that the S/N for the heterodyne case is 3 dB higher. That is, the coherent receiver will need half the power that a direct noncoherent receiver requires to achieve the same S/N performance. To repeat, it is important that the heterodyne receiver design maintain proper phase matching between the input signal and the local oscillator, as well as maintaining the polarization match. The mixing efficiency must also be high in performance.

The coherent receivers, based on the design of their local oscillator systems, may have the special advantage of being tuned to different frequencies, so that different wavelength signals can be transmitted to the coherent receiver from different signal source platforms.

The generic block diagram of the heterodyne receiver is shown in Figure 2.10 and another, a related coherent detection receiver known as the homodyne receiver, is shown in the schematic in Figure 2.11.

The homodyne receiver is more efficient by virtue of the fact that the pickoff for the feedback to drive the local oscillator of the heterodyne process is at the optical receiver load, R_L. The conceptual design of the heterodyne receiver shown in Figure 2.10 has its pickoff point past the IF amplifier.

In terms of the S/N ratio for the homodyne:

$$S/N = \left\{2\left(GQq/h\nu\right)^2 P_{LO}P_C R_1\right\} \div \{G^2 q\left[Qq/h\nu\left(P_{LO} + P_C + P_B\right) + I_D\right]$$
$$\times BRl + 2kTB\} = 2QP_C/h\nu B \text{ when } P_{LO} \gg P_B, P_C, 2kTB$$

The SNR of the homodyne is thus 4 times as large as that of the direct detection receiver.

A separate version of the heterodyne receiving system, which enhances its performance, consists of breaking out the local oscillator output wave into vertical and horizontal polarization and mixing each with the input signal wave, as shown in Figure 2.12. The separation of the vertical and horizontal polarization components of the signal entering the RF sections enables the optimum extraction of the input signal. However, this receiver is more complex than the direct detection receiver.

2.5.3   Other Modulation Formats

While the above two modulation schemes associated with the direct and the coherent receiver systems are the basic modulation formats, there are other modulations and associated receiver types with unique advantages. They are based on the environments in which the communication systems are to operate.

For example, for the direct detection model, use of OOK modulation has advantages over PGBM because the pulses are easier to generate and typically have a higher peak power level. Further, the receiver does not have to depend on the timing and blocking process. The envelope of the ON pulse, the *1*, encloses the optical frequency for the period of the *1* bit.

Shown in Figures 2.13 and 2.14 is the pulse polarization binary modulation (PPBM) and the pulse position modulation (PPM) system with

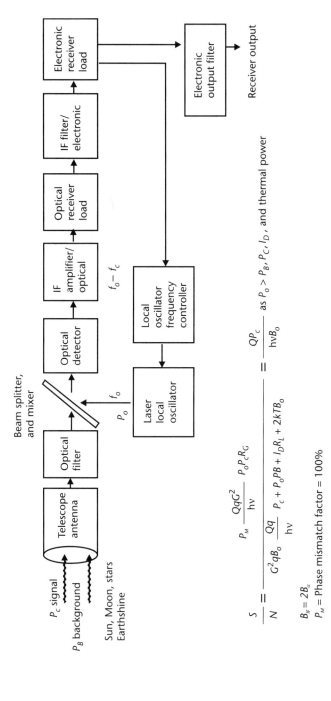

Figure 2.10 Generic block diagram of the heterodyne (coherent) detection receiver.

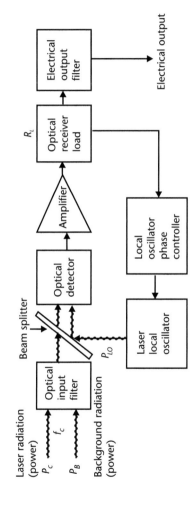

For $B_{if} = 2B_o$ synchronous detection

$$\frac{S}{N} = \frac{2QP_c}{h\nu B_o} \quad \text{for } P_o > P_{LO} \text{ and } P_c$$

Figure 2.11 The homodyne receiver, top view block diagram.

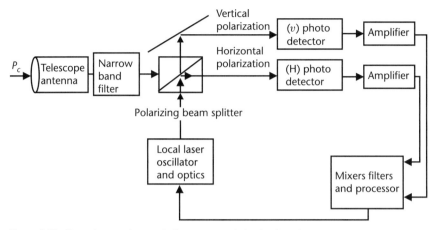

Figure 2.12 Heterodyne receiver employing separate polarization detection.

their waveform format. As shown, the PPBM has the advantage of inherent reliability in that the system always expects a pulse. It also has the special advantage of being operable in a noisy environment without substantially increasing its BER, when compared with other modulation schemes.

Shown in Figure 2.15 is the pulse position modulation (PPM), with its salient features summarized. This modulation has been recommended by NASA for various laser space communications including planetary missions. PPM has the unique property of enabling many bits to be transmitted by a single pulse. This is done by placing the pulse at a specified time slot position, between specified synchronization pulses.

In PPM, pulses are transmitted at equal intervals. However, as indicated in Figure 2.15, the synchronization pulses may be not be required to be transmitted; they may be generated in the receiver, providing the same system benefit. Moreover, they may be synchronized by an external source.

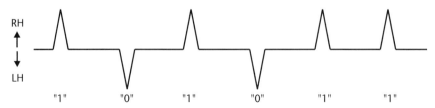

- Similar to PGBM except polarization is switched for laser pulse output
- Pulse is always expected at receiver
- Requires dual channel (polarized) receiver

Figure 2.13 Pulse polarization binary modulation format.

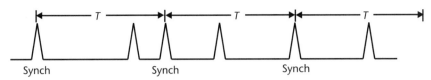

- Sampled analog pulse modulation format
- Many bits/pulse transmitted
- Direct analog-to-pulse position data conversion
- Direct reconversion from pulse position to analog
- Repetition rate must be at least twice the analog
  information bandwidth to meet sampling theory requirements.

Figure 2.14 Pulse position modulation.

Further shown in Figure 2.15 are examples of M intervals, the duty cycle, and bits/pulse.

The PPM implies a direct data conversion from analog to pulse position and direct reconversion from pulse position back to analog. However, the repetition rate must be at least twice the analog information bandwidth to meet the sampling theory requirements. In PPM, information is transmitted by sending a pulse in one of M possible time slots, each of duration (ΔT) seconds. Thus the data rate, symbolized by D_{rate}, is

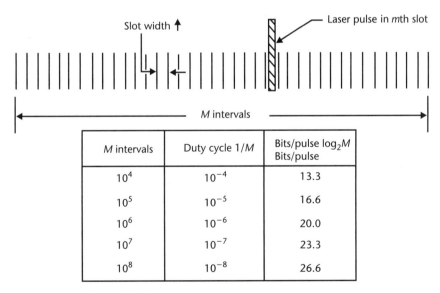

M intervals	Duty cycle 1/M	Bits/pulse $\log_2 M$ Bits/pulse
10^4	10^{-4}	13.3
10^5	10^{-5}	16.6
10^6	10^{-6}	20.0
10^7	10^{-7}	23.3
10^8	10^{-8}	26.6

- Many bit/pulse
- Requires low power pre-encoding electronics
- Potentially most efficient format

Figure 2.15 Pulse position modulation without explicit transmission of the synchronization pulses.

$$D_{rate} = Log_2 M / M(\Delta T) \qquad (2.22)$$

Defining α as log to the base 2 of M gives

$$\alpha = \log_2 M = \text{alpha level} \qquad (2.23)$$

where $M(\Delta T)$ is the number of seconds in which α is transmitted. Further, from Poisson statistics, the minimum required signal counts per decision interval are

$$S_{min} = -\log(2P_E) \qquad (2.24)$$

where P_E = bit error probability

The received Power, P_R, for the case of a laser communication system using PPM, may be expressed as

$$P_R = (D_{rate} / \alpha)(S/Q)(h\nu) \qquad (2.25)$$

where S = required signal counts per decision interval (photoelectrons per bit).

Substituting the above equations into the signal power budget first expressed in its simplest form, we have for P_T,

$$P_T = \{(4\pi R)(\lambda)/(\pi D_T)(\pi D_R)\}^2 \{(h\nu)(D_{rate})(S)/\alpha QF\} \quad (2.26)$$

2.5.4   Laser Communication Between Mars and Earth Using PPM Modulation

In the discussion to follow, calculations show that a laser link can be established between Mars and Earth. However, it should not be assumed that such a communications system will be implemented by NASA within the next few years, primarily because of cost issues.

Neglecting atmospheric losses, the following are the calculation components of the laser communication power budget, between a Mars station and an Earth-based station, which show feasibility:

R = distance between Mars and Earth stations, assumed for this calculation to be 240 million km ($2.4 \cdot 10^{11}$m). Clearly, the exact value is based on the orbital position trajectories of the two planets

D_T = 0.406m

D_R = 5.08m (~200 in)

λ = $0.53 \cdot 10^{-6}$m

$Q \cdot F \cong 0.2$ = (quantum efficiency) · (signal losses in transmitter and receiver)

where

$F = L_T \cdot L_R$
$\alpha = \log_2 M = 5$ alphabet level; bits per pulse
$\tau = 0.1$ pulse width in ms
S = signal counts per decision interval
$S_{min} = -\log_2\{2P_E\}$; for $P_E = 10(-3)$, $S_{min} = 6.2$
$D_{rate} = \log_2 M \div M(\Delta T) = 5 \div 148\{0.1 \cdot 10(-6)\} = 10$ Mb/sec

Using (2.26), we get $P_T = 1.44$ watts.

2.5.4.1   Estimated Data Rate [3]

A more conservative estimate of data rate for a laser signal between a transmitting station on Mars and an Earth-based ground station is 1 MBPS, when the optical aperture on Mars is 10 cm, its transmitter average power is 3W, the distance between the two stations is 1 AU, and the Earth's aperture is 5m. (The AU is the mean distance between the center of the Sun and the center of the Earth and is essentially 149,597,871 kilometers, that is, roughly 93 million miles.) At a distance of 2 AU, the data rate goes down to 0.1 MBPS, and it declines to 0.01 MBPS at 2.4 AU. Another configuration, which uses a 20-cm aperture on Mars and keeps all other system parameters the same, will achieve 10 MBPS at 1 AU, 1 MBPS at 2 AU, and 0.1 MBPS at 2.4 AU.

For a potential link design in which a satellite orbiting Mars, known as the Mars Reconnaissance Orbiter (MRO), could be used to communicate with Earth, the estimated data rate would be 70 MBPS at 1 AU when the output power is 3W of average power and the aperture is 30 cm on the satellite and 10m at the Earth. At a 2 AU distance, the data rate would be 10 MBPS; and at 2.4 AU, 8 MBPS.

2.6   Expression of Signal Power Budget Due to Vibrations

As shown in (2.6), two terms in the signal power budget indicated the loss of signal due to vibrations of the spatial platform and the noise in the acquisition and tracking circuitry. These are L_{P-T}, the pointing error of the transmitting beam, and L_{P-R}, the pointing error of the receiver optical beam. In evaluating these terms for the intersatellite link (ISL), one needs to consider the effect of the platforms' vibrations on the pointing loss of the signal beam and also of the beacon beam or a high-data-rate link going in the opposite direction, as would be the case in the transmit/receive functions of a laser satellite constellation. These vibrations cause the transmit beam to move away from the center of the receiver's telescope antenna. That is, the vibrations' amplitude and frequency that are superimposed on the beam cause it to fluctuate away from the center of the collector telescope antenna, resulting in an increase in the BER of the communications links.

2.6.1 The Pointing Loss Factors

The research of Chen and Gardner [4] and other investigators, Barry and Mecherly [5], Toyoshina et al. [6], and primarily Arnon, Kopeika, and their team [7, 8, 9, 10] have led to the detailed analysis of laser space communications performance in the face of the physical vibration environment, including the noise in the electro-optic tracking subsystem. For this discussion, we also borrowed from the engineering simulation work done in this area at Jet Propulsion Laboratory [11].

The sensitive alignment accuracy that is necessary between two articulating satellites is also dependent on the isolation that is designed and achieved onboard the satellites between the vibration due to the electro-optic tracker, the vibration induced by the satellites' mechanical components, and the inherent orbital motions of the satellite. These effects will result in the overlay of vibrations on the photon beam, which will lead to a reduction of the number of received photons. Particularly for the case of the heterodyne receiver, this will cause a reduction in the mixing efficiency, which will lead to a further increase in the bit error rate.

2.6.2 Mathematical Expressions for the Pointing Losses

It is first assumed that the pointing error angle, θ_R, centered along the radial line between the transmitting satellite and the receiving satellite, is made up of a steady state pointing error and a random pointing component of the pointing error. The latter, in turn, is composed of pointing angle error, along the azimuthal axis, known by θ_{az}, and a pointing error angle along the elevation axis, known as θ_{EL}. Each is assumed to be independent of the other, and each is normally distributed. This assumption is typically used in physical examples where random processes are involved.

The probability density function (PDF) of θ_{az} may then be written as

$$f\left(\theta_{az}\right) = 1/\left\{(2\pi)^{1/2}\,\sigma_{az}\right\}\exp-\left\{\left(\theta_{az} - \mu_{az}\right)^2/2\sigma_{az}^2\right\} \qquad (2.27)$$

where

    σ_{az} = standard deviation of the random component of the error angle along the azimuthal axis

μ_{az} = mean value of the random component of the error angle along the azimuthal axis

Based on the same assumptions it may be stated that the normal PDF of the random error angle along the elevation axis, $f(\theta_{EL})$, is equivalent to the PDF of the azimuthal angular error given in (2.27). Furthermore, because it is also reasonable to assume that the random error angle components along the azimuthal and the elevation axes are independent, we may define for the case of symmetry with no bias: The radial angle error squared is equal to the sum of the squares of the azimuthal and elevation error angles;

$$\theta_R^2 = \theta_{az}^2 + \theta_{EL}^2 \tag{2.28}$$

Based on symmetry, we can also express the variances of those error angles as

$$\sigma_R = \sigma_{az} = \sigma_{EL} \tag{2.29}$$

In the simplified mathematical model that is discussed in this section, emphasis is placed on clarifying the effect of the physical characteristics of the vibration phenomena, together with chosen modulation schemes, to show performance parameters of the system under vibration. This, rather than a detailed analysis of the actual system design, is considered appropriate here. This approach helps to elucidate the overall system's performance and evaluate its sensitivity to different inputs.

From the previous equations, for the case of zero bias, we get the Rayleigh distribution function for the transmitter and receiver pointing error angles, in the jitter environment:

$$f(\theta_T) = \theta_T / \sigma_T^2 \exp\left\{-\theta_T^2 / 2\sigma_T^2\right\} \tag{2.30}$$

$$f(\theta_R) = \theta_R / \sigma_R^2 \exp\left\{-\theta_R^2 / 2\sigma_R^2\right\} \tag{2.31}$$

2.6.3 Satellite Vibrations and Their Effect on the Communication Link Performance for a Typical Laser Transceiver Design for ISL

Before the BER is calculated for the laser transceiver in an ISL application, an evaluation of the losses due to the transmitter beam pointing and the receiver beam pointing, which are factors in the SPB, will be described.

The SPB relationship given in (2.6) is restated for convenience and is now called (2.32):

$$n' = P_T L_T G_T G_R L_R Q \left(L_{P-T}\right)\left(L_{P-R}\right)/L_S\left(h\nu\right)f \qquad (2.32)$$

The key parameters effecting the pointing error are the optical antenna gains $G_T = (\pi D_T/\lambda)^2$ and $G_R = (\pi D_R/\lambda)^2$ and also the θ_T^2 and θ_R^2, the square of the radial pointing error of the transmitted beam, and the square of the radial pointing error of the receiver beam, respectively. The signal losses due to incorrect pointing are then given by (2.33) and (2.34), provided the beams' cross-sections are Gaussian:

$$L_{P-T} = \exp\left(-G_T \theta_T^2\right) = \exp\left\{-\left(\pi D_T / \lambda\right)^2 \theta_T^2\right\} \qquad (2.33)$$

$$L_{P-R} = \exp\left(-G_R \theta_R^2\right) = \exp\left\{-\left(\pi D_R / \lambda\right)^2 \theta_R^2\right\} \qquad (2.34)$$

Substituting these two equations into (2.32) yields, for the case of a laser transceiver having a single telescope to transmit and receive (i.e., $G_T = G_R$ coupled through a circulator)

$$n' = P_T L_T G^2 L R Q \exp\left\{-G\left(\theta_T^2 + \theta_R^2\right)\right\}/L_S\left(h\nu\right)f \qquad (2.35)$$

The plot of the BER as a function of the ratio of σ, the root-mean-square (RMS) of the amplitude of the vibration to the laser beamwidth, for the OOK modulation, has been developed by Arnon [5]. It is presented here as Figure 2.16.

In Figure 2.16, it is shown that when the RMS of the amplitude of the pointing jitter is $\leq 7\%$ of the transmitter beamwidth λ/D_T, then the BER will be no greater than $\sim 10^{-9}$. However, should the jitter amplitude exceed $\leq 7\%$, for example, from 0.07 to 0.10 of the beamwidth, then the error rate will then leap from 10^{-9} to more than 10^{-5}.

The receiver used in this direct detection example includes an optical preamplifier that manifests, apart from signal shot noise, background shot noise, dark current noise, thermal noise, and the noises created by the amplified spontaneous emission (ASE).

2.6.4   Effect of Vibration on the Communication Links Through a Constellation of Satellites

Constellations of low-altitude satellites, are the space segment of worldwide cellular links (colloquially described as "wireless" or "fiberless" networks) enabling communication between a ground station in one location of the Earth and a ground station in another part of the Earth. Conceptual space systems such as the Teledesic and Iridium are good examples of potential space segments. Such constellation networks may be composed, for example, of κ

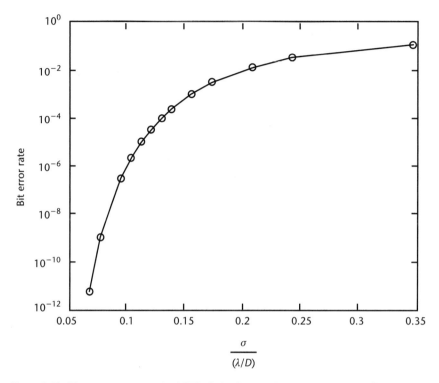

Figure 2.16 Bit error rate versus ratio of RMS of vibration intensity to laser beamwidth. [7]

satellites. However the error rate of the data stream could just "pile on" and accumulate, as a result of the relaying (or repeating) function, from one satellite to the next. If one were to demodulate the received data and apply error detection and correction software and then retransmit the data signal to the next satellite, there would clearly be a corresponding reduction in error rate.

However, for a planned low-cost spatial network system, only the relaying function—the amplification—is considered; therefore, errors may accumulate and build up. Consequently, considerable care must be applied to achieve a high degree of isolation from the mechanical vibration and separation from the electrical noise of the tracking circuit and the laser pointing error. Clearly, an adequate signal margin must be included.

2.7 Azimuthal and Elevation Components of the Pointing Error Angle in a Constellation of Satellites

Based on the normal distribution, the azimuthal pointing error angle of the ith satellite in the constellation may be expressed as

$$f\left(\theta_{az}-i\right)=1/\left\{(2\pi)^{1/2}\left(\sigma_{az}-i\right)\right\}\exp\left\{-\left(\theta_{az}-i\right)^2/2\left(\sigma_{az}-i\right)^2\right\} \quad (2.36)$$

where $\sigma_{az}-i$ = RMS of the azimuthal pointing error angle, of the ith satellite
Furthermore,

$$f\left(\theta_{EL}-i\right)=1/\left\{(2\pi)^{1/2}\left(\sigma_{EL}-i\right)\right\}\exp\left\{-\left(\theta_{EL}-i\right)^2/2\left(\sigma_{EL}-i\right)^2\right\} \quad (2.37)$$

where $\sigma_{EL}-i$ = RMS of the elevation pointing error angle of the ith satellite

The radial pointing error angle of the ith satellite may be expressed as the root mean square of the azimuthal and elevation angular components, as follows:

$$\theta_R-i=\left\{\left(\theta_{az}-i\right)^2+\left(\theta_{EL}-i\right)^2\right\}^{1/2} \quad (2.38)$$

With the same symmetry assumption as used in a single pair of articulating satellites, we can describe the ith satellite in the constellation net as

$$\sigma_R-i=\sigma_{az}-i=\sigma_{EL}-i \quad (2.39)$$

Then, based on the independence of each of the two variables and their identical distribution, the distribution of the radial pointing error angle of each is Rayleigh:

$$f\left(\theta_R-i\right)=\left(\theta_R-i\right)^2/\left(\sigma_R-i\right)^2\exp\left\{-\left(\theta_R-i\right)^2/2\left(\sigma_R-i\right)^2\right\} \quad (2.40)$$

The expression for S/N ratio between the ith receiver satellite and a transmitter satellite when receiving a "1" in the OOK modulation may now be written as

$$\begin{aligned}SNRi-OOK-1=\int\left(d\theta_R-i\right)S\left\{\left(\theta_R-i-OOK-1\right)\right\}^2\\f\left(\theta_R-i\right)/\left\{\left(\theta_R-i\right)OOK-1\right\}^2\end{aligned} \quad (2.41)$$

where $S(\theta_R-i)$ = $P_T\{qQ/h\nu\}\{L_TL_R/L_S\}G_TG_RL\tau$.

We can also write $L\tau$ = the loss of transmitter signal due to pointing error as

$$\exp(-G_T\theta_T^2)$$

When "1" is not transmitted in the OOK modulation, then a "0" is assumed to be the transmitted bit.

Having the SNR relation for the ith satellite, we can now write the relationship for the BER for the entire spatial circuit composed of κ satellites, with the pointing error angle in the radial direction, $\theta_R - i$, for the ith satellite. We start with the instantaneous probability of error for the OOK modulation for the same ith satellite, using the complementary error function, which is defined as

$$\mathrm{erfc}(x) = 1 - 2/\sqrt{\pi} \int \left\{ \exp\left(-y^2\right) dy \right\}$$

We may then state, as in Polishuk and Arnon [2.7], the probability of error for the OOK data stream modulation:

$$P_{OOK}\left(\theta_R - i\right) = \alpha\,\mathrm{erfc}\left\{ \frac{\left[\left(S(\theta_R - i)_{OOK} - 1\right) - \left(S(\theta_R - i)_{OOK} - 0\right)\right] \div}{\sqrt{2}\left[\sigma(\theta_R - i)_{OOK} - 1 + \sigma(\theta_R - i)_{OOK} - 0\right]} \right\} \quad (2.42)$$

where the "1" is the ON signal and "0" is the off signal in the OOK signal modulation.

The BER of the ith satellite in the entire circuit of satellites in the constellation, for OOK data modulation, may be written as

$$BER_{OOK} - 1 = \int_0^\infty P_{OOK}\left(\theta_R - 1\right) f\left(\theta_R - i\right) d\theta_R - i \quad (2.43)$$

The BER for the entire circuit of satellites may then be expressed by multiplying (2.43) by as many satellites as there are in the network between the two articulating nodes. That is, the signal is transferred from the initial satellite, which receives an uplink message from a ground station (the caller), and gets connected to a number of satellites acting as a relays nodes, until the satellite transmits downlink to the nearest ground station of the receiver station (the callee). Thus, the BER of this network is multiplied by the error rates of the spatial platforms from i = 1 to i = κ. This expression may be stated as

$$BER_{network} \cong 1 - \Pi\left[1 - BER_{OOK} - i\right] \quad (2.44)$$

When $BER_{OOK} - i$ is much less than 1 for all satellites in the circuit, then (2.44) reduces to the sum, Σ from i = 1 to i = κ

$$BER_{network} \cong \Sigma\left[BER_{OOK} - i\right] \quad (2.45)$$

Clearly, if all the κ satellites are equal, the $BER_{network}$ will be equal to approximately κ times the BER_{OOK} of any one of the satellites. And if one satellite has a high BER relative to all the others, then $BER_{network}$ is approximately equal to that satellite BER_{OOK}. This potential emphasizes the need to make sure that all BER due to pointing errors be very small. The error rate can be significantly reduced by providing sufficient isolation between the mechanical vibrating elements on the spatial platforms and the noise on the tracking systems and also by improving the signal margin between satellites. Also to be addressed is improving the quality and reliability of the active and passive components and subsystems of the laser transceivers in the network.

2.8 Summary and Concluding Remarks

This chapter reviews the calculations necessary to evaluate the SPB between spatial platforms, such as two synchronous satellites and between low-altitude satellites and synchronous satellites (and by extrapolation, between a satellite and an airborne platform). A method of determining the BET for a given link with a particular modulation scheme, such as PGBM, is shown. Other modulation schemes are outlined, and overview schematics of a direct noncoherent receiver and coherent (heterodyne) receiver are presented.

Calculations are also presented for the conceptual link between Mars and Earth at distances of 2 AU and 3 AU. In such links, the data rates are of the order of several megabits per second.

Attention is directed to the evaluation of the BER as a function of the radial pointing error angle. This radial angle error is comprised of the RMS of the azimuthal and elevation angular components of the angle error. It is shown that BER will remain low ($\sim 10^{-9}$) if the jitter of the beam $\leq 7\%$ of the beamwidth λ/D. If, however, the jitter goes up to 10% of the transmitted beamwidth, the BER jumps to 10^{-5}. The considered modulation scheme is OOK, a form of PGBM.

Finally, it is also shown that for a constellation of satellites, if every one of satellites acts as a "repeater," then the error of each link just piles on and the total BER at the callee's end will be substantial. The methods of avoiding this situation involve employing an error detection and correction modality (at each node), as well as correcting and compensating for the platform vibration by the technique indicated in Chapter 3.

References

[1]   Ross, M., P. Freedman, J. Abernathy, G. Matassou, J. Wolf, and J. D. Barry, "Spatial Optical Communications with the Nd:YAG Laser, Proc. IEEE Vol. 66, No. 3, March 1978.

[2]   Pratt, W. K., *Laser Space Communications*, New York: John Wiley & Sons 1971.

[3]   Biswas, A., and S. Piazzola, "Deep-Space Optical Communications Downlink Budget from Mars: System Parameters," *JPL-IPN Progress Report*, 42–154, 2003.

[4]   Chen, C. C., and C. S. Gardner, "Impact of Random Pointing and Tracking Error on the Design of Coherent and Non Coherent Optical Intersatelliote Communications," *IEEE Trans. on Communication*, Vol. 37, 1981, pp. 252–260.

[5]   Barry, J. D., and G. S. Mecherle, "Beam Pointing Error Significant Design Parameters for Satellite-born, Free Space Optimal Communication Systems," *Optical Engineering*, Vol. 24, pp. 1049–1054.

[6]   Toyoshina, M., T. Jono, K. Nakagawa, and A. Yamamoto, "Optimum Divergence Angles of a Gaussian Beam Wave in the Presence of Random Jitter in Free Space Communication Systems, *Journal of the Optical Society of America*, Vol. A19, 2002, pp. 567–571.

[7]   Polishuk, A., and S. Arnon, "Optimization of a Laser Satellite Communication System with an Optical Preamplifier," *Journal of the Optical Society of America*, Vol. 21, No. 7, 2004.

[8]   Arnon, S., S. R. Rotman, and N. S. Kopeika, "Performance Limitations of Free-Space Optical Communication Satellite Networks Due to Vibrations: Direct Detection Digital Mode," *SPIE/Optical Engineering*, Vol. 36, No. 11, 1997, pp. 3148–3157.

[9]   Arnon, S., S. R. Rotman, and N. S, Kopeika, "Beamwidth and Transmitter Power Adaptive to Tracking System Performance for Free-Space Optical Communication," *Applied Optics*, Vol. 36, 1997, pp. 605–610.

[10]   Arnon, S., S. R. Rotman, and N. S. Kopeika, "Bandwidth Maximization for Satellite Laser Communication," *IEEE Trans. on Aerospace and Electronic Systems*, Vol. 35, No. 2, 1999.

[11]   Lee, S., G. G. Ortiz, J. W. Alexander, A. Portillo, and C. Jeppesen, "Accelerometer-Assisted Tracking and Pointing for Deep Space Optical Communications: Concept, Analysis, and Implementations," IEEE Aerospace Conference, Big Sky, MT, 2001.

3

Acquisition Tracking and Pointing

Arthur Kraemer and David Aviv

3.1 Introduction

In the previous chapter, we primarily developed expressions for the SPB between two spatial platforms supporting intersatellite links at a synchronous altitude and both at low altitudes. In this chapter, we go into the methods of acquiring and tracking of the two platforms with one at low altitude and one at synchronous altitude. Chapter 3 reviews the work that has been done in the area of acquisition, tracking, and pointing (ATP) between satellite platforms, primarily in the United States and supported by NASA and the U.S. Air Force/Department of Defense (USAF/DoD). Almost all of NASA's work involves linkages from a satellite or spacecraft to Earth. Other U.S. work has planned laser communication links between synchronous satellites and synchronous-to-ground links. Such systems involved a combination of unique environmental satellite applications and transformational communication systems. The Europeans, however, have undertaken, through their SILEX Project, low-Earth-orbit (LEO) satellite to geosynchronous-equatorial-orbit (GEO) satellite and the implementation of laser communications between the GEO and an Earth station.

The analysis of the ATP system starts with a description of the point-ahead angle (PAA). The initial pointing is based on determining the position of the two satellites, followed by the process of each tracking the other. Particularly in the case of an LEO satellite, its position is determined by virtue of the GPS; the GPS satellites are at ~20,000 km in altitude, and the PAA is typically calculated by the ATP system of the GEO satellite. The GEO will direct its beacon laser to be ahead of the LEO, so that when the LEO travels along its trajectory, it will intercept the beam and thus enable the closing of the communication link between the satellites.

The discussion in this chapter next proceeds with a pictorial description of the three phases of the acquisition, pointing, and tracking between the LEO and the GEO satellites. With a sensor suite system assumed to be deployed on the LEO, the resulting high-data-rate signal will be used to modulate the laser beam and transmit the signal to the GEO. The latter may relay the data to another satellite for retransmission to a selected ground station, or the GEO can retransmit the sensor suite data directly to another ground station or an airborne platform.

Several block diagrams of typical transceivers onboard the satellite platforms are presented to provide an overall understanding of each of the subsystems. From these diagrams, it is recognized that in the GEO's ATP system, the beacon laser may need to scan the region of location uncertainty of the LEO. (The location and attitude of the LEO is known within a region of a small uncertainty value.) A mathematical expression, initially developed by Arthur Kraemer [1] for the scanning procedure, provides an analytical approach for the scan design. A comparison is made with an alternate approach; the scanned beam versus the broadened beam design. Both designs are sensitive to the background radiation, due to the Sun, Moon, and other spectral radiation sources that may enter the field of view (FOV) of the telescope. This telescope is located on the LEO. It searches upward for the GEO platform, on which the Beacon source resides.

Several approaches have been developed to improve the tracking process, in the face of physical vibration of the satellite subsystems and components, which affects the pointing of the laser beam. Accelerometers and other inertial instruments are used to detect and process the physical vibrations. The equivalent double integration of the output of the accelerometer is used to achieve a position plot, which is then used to develop an error signal, relative to the reference curve gotten from the focal plane array (FPA). The error signal is used to servo the movement of the fine scanning mirror (FSM) and, as necessary, to make corrections of the coarse gimbal movements as well. The fine and the coarse subsystems will be slaved to one another.

Major work in this area has been done at Jet Propulsion Lab (JPL), [2, 3], where mathematical simulations, software development, and laboratory experiments in the field of vibration compensation systems for satellites have been performed. Associated with the ATP design is a special subsystem originated by M. Fitzmaurice of NASA/Goddard [4] and improved on by several investigators at MIT Lincoln Labs and JPL. The system is composed of an FPA composed of a two-dimensional CCD array, for the determination of the error signal. A signal derived from a small fraction of the high-data-rate laser transmitter is considered the reference, while the second signal, displayed on the same FPA, is derived from the position of the FSM. The resulting difference, the error signal, is used to adjust the position of the FSM. The servo loop bandwidth is of the order of several thousand Hz. That is fast enough to accommodate any small changes in the beam directions due to the vibration field.

A second FPA may also be used to adjust the position of the slow or coarse mirror, or adjust the gimbals of the telescope antenna. The closed loop bandwidth of this servo is of the order of 100 Hz or less. Apart from the direction of the angle of the beam arrival, the coarse- and fine-pointing corrections are obtained from the accelerometers' subsystems.

In some ATP system architectures, compensation of the vibrations that are superimposed on the optical system can also be achieved by detecting the beam arrival angle and using this data as the input signal to the compensation servo system. For example, the beam may be tilted to one side instead of the center of the detector array or to the apex of the pyramidal four-corner cube. The derived error signal from the quadrant detector subsystem may then be used to correct the coarse- and fine-steering mirrors' positions.

As was discussed in Section 2.6 and Section 2.7, without these techniques of compensation for the vibration, the deleterious results of these effects on the BER could be severe. Moreover, the analysis developed by a number of investigators in the vibration field may also be useful in estimating the degradation of performance of the communications links, in the event of failure of one or more components of the ATP subsystems.

3.2 Implementation of the ATP Functions

Accurate implementation of the ATP functions between articulating satellites is required in order to enable the flow of data between the spatial platforms and between satellites and ground stations. In this chapter, examples of feasible designs of the ATP between a LEO and a GEO satellite are described, and key details are presented. The example starts with the PAA and is followed by a discussion of the use of a beacon laser, which emits its beam from the GEO toward the LEO. This is followed by the LEO's acquisition of the beacon beam, which it does by tracking it and then responding to the GEO platform, by transmitting a wideband data beam toward it.

The downlink/uplink between a satellite and a ground station, which also requires an ATP process, is further complicated by the fact that the beacon and communication beams have to go through the atmosphere, requiring adjustments due to turbulence and accommodating when possible for the deleterious effects of absorption, scatter, and weather issues. Those losses are estimated and methods of amelioration are discussed in Chapters 4, 5, and 6.

3.2.1   Functional Description

A generalized acquisition sequence between the LEO and GEO platforms is shown in Figures 3.1, 3.2, 3.3, and 3.4.

An acquisition laser beacon on the GEO, directed toward the LEO, is either spread or scanned over the full attitude uncertainty (±0.2°) of the LEO. The beacon is aimed with the aid of the PAA data. The latter is calculated at the GEO, and the result is used to direct the beam toward the anticipated orbital position of the LEO satellite.

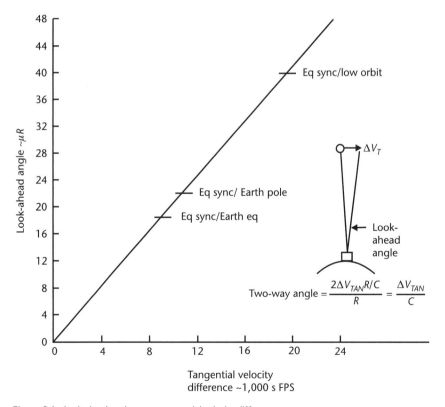

Figure 3.1 Look-ahead angle versus tangential velocity difference.

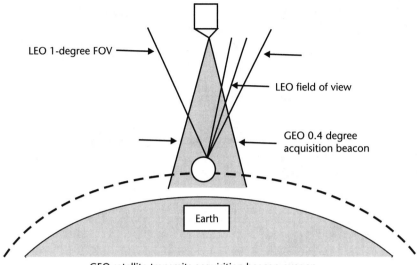

LEO 1-degree FOV

LEO field of view

GEO 0.4 degree
acquisition beacon

Earth

GEO satellite transmits acquisition beacon over an
angle equal to its attitude uncertainty of 0.4 degrees.

LEO searches its field of view over its attitude
uncertainty of 1 degree to locate beacon.

Figure 3.2 Acquisition Phase I.

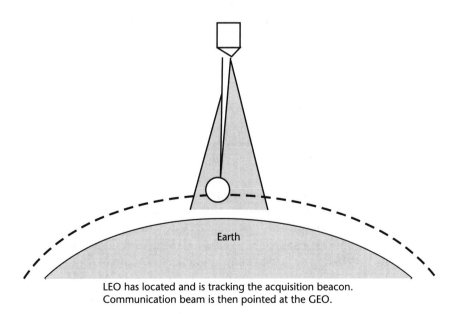

Earth

LEO has located and is tracking the acquisition beacon.
Communication beam is then pointed at the GEO.

Figure 3.3 Acquisition Phase II.

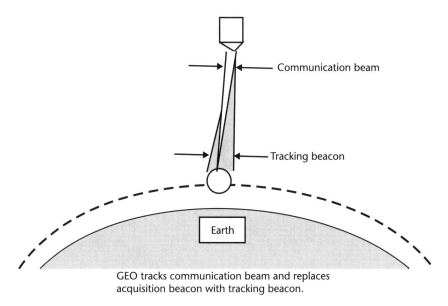

GEO tracks communication beam and replaces
acquisition beacon with tracking beacon.

Figure 3.4   Acquisition Phase III.

The PAA configuration is shown in Figure 3.1. The two-way angle is equal to twice the tangential velocity differential, ΔV_{tan}, between the two satellite platforms, times the time, R/C, for the photons to get there. That is,

$$2\text{-way angle} = 2\Delta V_{tan}(R/C)R = 2\Delta V_{tan}/C \qquad (3.1)$$

Or equivalently,

$$\sin \theta = \Delta V_{tan}/C$$

Continuing with the four major steps in the acquisition and tracking process between the synchronous and the low altitude satellites,

1. At a range of about 40,000 km between the GEO and the LEO, the beacon will cover an area that is 280 km in diameter. Without the GPS, the position of the LEO can be determined from its ephemeris to an accuracy of about 10 km (with GPS and certainly with differential GPS, the position accuracy of the LEO satellite can be of the order of meters), so that there is large confidence that the beacon will capture the LEO. However, during this search time period, the LEO searches an FOV that corresponds to the

attitude uncertainty of ±0.5 degree until the beacon beam is located (Figure 3.2).

2. The LEO then goes into a tracking mode, turns on its laser communication subsystem, and directs the data stream to the GEO (Figure 3.3).

3. The optical antenna telescope of the GEO searches its FOV for the upward-directed data beam. It initiates a tracking procedure and directs a narrow tracking beacon beam to the LEO (Figure 3.4).

4. The LEO locks on the tracking beacon and accurately points the data beam to the GEO optical antenna, and the link closes. The high-data-rate signal then begins to flow to the GEO.

There are several variations of the basic steps just described. The beacon could be located on the LEO, and it may spread over the uncertainty volume of the GEO or scan that region with a narrower beacon beam. An alternate solution might be to initially use the communication beam as the acquisition beacon by either spreading ("spoiling") it into a wide angle or scanning it. However, the beacon laser should have a high peak power and lower pulse rate (~100 pulses/sec) to help the beacon receiver locate the beacon beam in the presence of background radiation coming from the Earth, the Moon, or even the Sun. A high-peak-power laser requires a high-gain laser cavity (unless a complex array of diode lasers is used), while the communication laser is most likely to be a low-gain laser operated with a continuous output power. Therefore, the communication laser does not make a very good beacon, particularly if its beam is spread. It can be tracked by the satellite containing the communication receiver to maintain pointing after the acquisition sequence has been completed and the communication laser is pointing directly at the receiver, as the communication transmitter on the LEO tracks the beacon laser.

Under these circumstances, if the LEO satellite were to initiate acquisition, it would be required to carry two lasers rather than just the communication laser. Alternately, the GEO satellite can carry a beacon laser that can be used to initiate ATP with several GEO satellites. All things considered, it makes sense for the GEO satellite to initiate ATP.

3.3 Basic Block Diagram of the ATP on LEO and GEO

The basic components of the laser communication system onboard the LEO and GEO is shown in Figure 3.5.

The LEO package contains the high-data-rate transmitter with bore-sighting alignment and PPA compensators in its optical path. The bore-sight

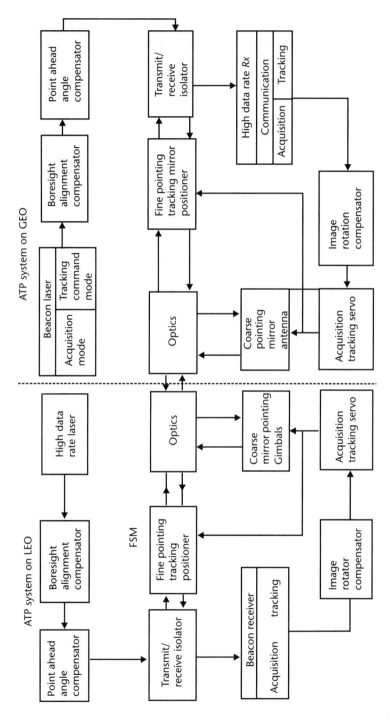

Figure 3.5 Block diagram of ATP SYSTEM Onboard LEO and GEO.

alignment is checked and adjusted with the transmitter and the beacon receiver before the acquisition phase. The look ahead compensates for the relative orbital motions of the two platforms in pointing the narrow communication beam. For the LEO to GEO, the PAA may be up to 74 μm.

The transmit/receive beam isolator (the circulator) is necessary, to avoid reflecting any of the output power back into the receiver. The beacon receiver has to acquire and track. During acquisition, the received beam is positioned by the coarse-pointing mirror to the narrow tracking field. Error signals from the tracking detector are transmitted by the acquisition and tracking servo into the mechanical motion of fine tuning and the tracking mirror positioner. The positioner, also called the fine-steering mirror (FSM), provides the very accurate pointing and tracking necessary for the narrow beam communication. It is to be pointed out, however, that recent technology developments have shown that the mechanical gimbal movements used in directing the mirrors can be replaced by the use of beam benders and Bragg reflectors, so that the beams move, rather than any mechanical orientation of the mirrors occurring [6].

To continue with the existing technology of mechanical movements of mirrors, the image rotation due to the azimuth motion of the coarse-pointing mirror must be compensated for between the receiver and the servo. Also, there is the additional design requirement during the tracking that command data may be received from the GEO and will need to be acted on.

The optics of the beacon receiver and the high-data-rate transmitter may be the same if the wavelengths are sufficiently separate, for example, 0.53–0.60μ for the beacon and 1.06–1.55μ for the high-data-rate transmitter signal. The coarse-pointing mirror keeps the communication beam pointed at the GEO for any relative satellite orbital positions, as well as for performing the acquisition positioning function.

The GEO satellite contains the high-data-rate receiver, which consists of separate communication, acquisition, and tracking detectors. The beacon transmitter consists of a laser that works in two modes: acquisition, a mode of very short duration (20 sec), and tracking, a mode of long time duration, during which command data may be sent. The bore-sight alignment compensator, point-ahead compensator, and transmit/receive isolator perform the same functions as in the LEO satellite.

The high-data-rate receiver has three functions: first, to receive the high-data-rate signal information; second, the acquisition and tracking functions that keep the narrow communication beam on the communication-detector while tracking; and third, to provide accurate pointing information for the beacon laser.

Error signals from the last two detectors are translated by the acquisition and tracking servo to the coarse-pointing mirror and fine-steering mirror, after compensation for image rotation. As in the LEO, the optics of the GEO may be common or separate. The GEO receiver primary is large and close to diffraction limited, while the beacon laser primary aperture is much smaller and may require beamwidth control. The coarse-pointing mirror antenna serves the same purpose as that of the LEO.

3.4 Specific Acquisition Procedures in Step 1

To more fully illustrate the acquisition procedure presented in the previous section, two possible implementations are considered next.

The beacon beam may be designed to search for the LEO by two different methods: generating a broad beam to cover the entire region of location uncertainty of the LEO satellite with a high-peak-power, low-repetition-rate laser, and scanning the entire region of uncertainty with a narrow beam having a high-repetition-rate laser with each pulse being of low power.

Figure 3.6(a) shows the broadbeam case and Figure 3.6(b) shows the narrow beam scan case. In each case, the LEO beacon receiver receives a pulse every $1/N$ sec. If in both cases the laser transmitter has the same average power, P_{ave}, and the beacon receiver at the LEO has the same aperture size, the received energy per pulse will be the same. This may be illustrated as follows: in the broadbeam method, the emitted energy per pulse is given by

$$E_B = P_{ave}/N \qquad (3.2)$$

where

E_B = energy of the photon per pulse, transmitted by the beacon beam
N = number of pulses per second
P_{ave} = average power of the beacon laser beam

For the broadbeam case, the received energy per pulse, E_{BR}, is given by

$$E_{BR} = \frac{E_B \cdot D_R^2}{\theta\mu^2 \cdot R^2} = \frac{P_{ave} \cdot D_R^2}{\theta\mu^2 \cdot NR^2} \qquad (3.3)$$

where

D_R = diameter of the beacon receiver aperture

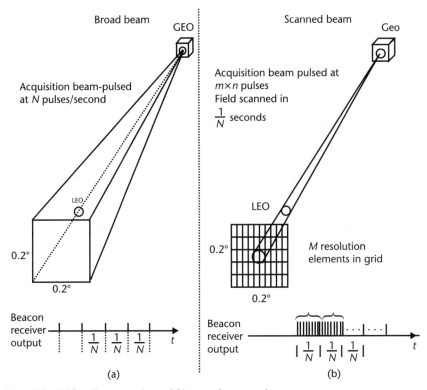

Figure 3.6 (a) Broadbeam scanning; and (b) narrow-beam scanning.

$\theta\mu$ = beacon divergence angle, equal to the full satellite attitude uncertainty of $\pm 0.2°$ $(0.4°)$

R = range of 40,000 km

In the scanned-beam case, the transmitted energy per pulse, E_S, is

$$E_S = \frac{P_{ave}}{M \cdot N} \tag{3.4}$$

where

P_{ave} = average power of the beacon beam
M = number of pulses in 1/N sec
$M \cdot N$ = total number of pulses/sec

The received energy per pulse for the narrow beam, E_{SR}, with beamwidth, θ_S, may be expressed as

$$E_{SR} = \frac{P_{ave}\, D_R^2}{\left(\theta_S^2\right)(M \cdot N)R^2} \qquad (3.5)$$

However, $\theta_S^2 = \theta\mu^2/M$ and $E_{SR} = E_{BR}$; therefore, both methods appear identical at the LEO's beacon receiver with respect to pulse rate and received photon energy.

There are significant differences in the performance of the two methods when considering the background noise photons. When the LEO beacon receiver optical antenna looks upward toward the GEO satellite, the Sun may be in the FOV of the antenna and can insert significant interference; therefore, special filtering must be used to select the major lines of the beacon laser, while blocking out much of the Sun's interfering spectrum.

With the Moon in the FOV of the LEO laser telescope antenna, its effect can also be reduced. This can be shown in the following way. The peak power received at the beacon receiver is as follows:

$$P_R = \frac{P_{ave}\left(D_R^2\right) T_O}{N\Delta\tau\theta_T^2 t_R^2} \qquad (3.6)$$

where

P_R = peak power received by the photodetector at the beacon receiver
θ_T = beamwidth of the beacon laser emanating from the GEO
N = pulse rate
$\Delta\tau$ = pulse width
T_O = optical efficiency of beacon receiver

The peak beacon current, I_{PR}, received at the LEO is given by (3.6), which is the same as (2.15) of Chapter 2:

$$I_{PR} = \left(Q\, P_R\, q\right)/h\nu \qquad (3.7)$$

where

P_R = peak power received in the photo detector
Q = quantum efficiency of the detector
q = charge of the electron
$h\nu$ = energy of a photon at frequency ν

The direct current, I_B, generated by a constant background source such as the Moon with spectral radiance N_λ, in units of watts/(cm)/ster A (where wavelength A is in units Angstrom) is given by

$$I_B = \left\{ \left(\pi^2 \right) Q N_\lambda \left(D_R^2 \right) \left(\theta_R^2 \right) \Delta\lambda \left(T_O \right) q \right\} / \{h\nu\} \tag{3.8}$$

where

D_R = beacon optical antenna diameter
T_O = optical efficiency of the beacon receiver
N_λ = Moon's spectral density
$\Delta\lambda$ = wavelength band of an optical bandwidth of the beacon receiver
θ_R = beamwidth of the beacon receiver FOV

The shot noise, also discussed in Chapter 2, is given by

$$I_N^2 = 2q_{IB}\Delta f \tag{3.9}$$

where Δf is the electrical bandwidth of a filter matched to the laser pulse-width, which can be closely approximated by $0.4/\Delta\tau$.

Now the signal-to-noise ratio for the detection of the acquisition pulses is, then,

$$S/N = I_{PR}/I_N \tag{3.10}$$

By substitution of (3.7), (3.8), (3.9), and (3.10), we have

$$\theta_R = \left[\frac{4\,P_{ave}D_R}{\pi N(S/N)\left(\theta_T^2\right)\left(R^2\right)} \right] \left[\frac{QT_O}{0.8(h\nu)\Delta\tau(N_\lambda)\Delta\lambda} \right]^{1/2} \tag{3.11}$$

To reduce the time required for the beacon receiver to acquire the transmitted beacon beam, it is important to make θ_R as large as is feasible, to the full attitude uncertainty of the LEO. To do this, it is necessary to maximize the terms in the nominator and minimize the terms in the denominator of (3.11), to the extent that it is consistent with realistic design limits of each of the parameters. As an example, let us consider the broadbeam case where the laser system is 2xNd:YAG, producing 0.53 μm (or Nd:YAG at 1.064 μm or laserdiode at 0.850 μm), and the beacon receiver uses a quadrant photomultiplier.

As seen from the typical parameters given in Table 3.1, there is very little flexibility in the beacon link design. For example, you could not maintain a link if the S/N were less than 10 dB and you could not make D_R exceed a reasonable size and weight on the LEO. Similar arguments can be made for the P_{ave}. After all, the advantage of laser over RF satellites is that of lower power, size, and weight. There is a limit as to the N, which cannot be made much smaller than 10 pps, and of course the N_λ of the Moon is fixed.

Table 3.1

Design Parameters for the Broad Beam Beacon Acquisition System

θ_T = 7 mrad = full GEO satellite attitude uncertainty
R = 40,000 km
Q = 20%
T_0 = 30%
$h\nu$ = 3.76 × 10 (-19)J/photons
N_λ = 4.7 × 10^{-7}W/cm^2 - ster A
$\Delta\lambda$ = 10A
S/N= 10
$\Delta\tau$ = 10^{-8} sec

Moreover, N, the number of pulses per second, must be large enough to enable the communication flow to be maintained. For example, if one were to choose N = 1 pulse/sec, it would very likely be that the communication link would break off. In fact, the minimum value of N is governed by the requirement of the LEO establishing a track on the Beacon downlink before the high-data-rate transmitter is turned on.

A servo loop operating on a series of N pulses would require about 50 pulses to reach equilibrium tracking conditions after the first pulse was detected. If, for example, 5 sec is an acceptable upper limit on the time to achieve the equilibrium tracking condition, the minimum allowable N would be 10 pulses/sec. Using N = 10 and the values in Table 3.1, we can write, for the broadbeam case,

$$\theta_R = 3.35 \cdot 10^{-2} P_{ave} D_R \qquad (3.12)$$

where θ_R, the allowable beacon receiver FOV is in radians, P_{ave} is in watts, and D_R is in centimeters.

Following a similar procedure, we obtain, for the scanned-beam case,

$$\theta_R = 6.32 \cdot 10^{-3} P_{ave} D_R \qquad (3.13)$$

In the acquisition case only, a CW pumped Q-switched laser may be used. The same parameter values are used except for $\Delta\tau$, for which this type of laser (2xNd:YAG) has a minimum of 2.8 · 10 (–7) sec.

In Figure 3.7, θ_R is plotted against D_R with P_{ave} as a parameter for both the broadbeam and the scanned-beam cases. If both have equal values of P_{ave} and D_R, θ_R can be approximately five times larger for the broadbeam case than the narrow-beam case.

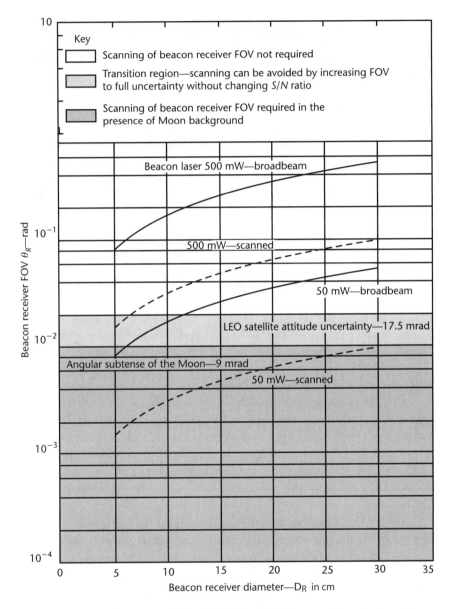

Figure 3.7 Maximum beacon receiver FOV to obtain S/N = 10 in the presence of moon background versus beacon receiver diameter.

Values of P_{ave} are constrained by the prime power on the GEO, and D_R is constrained by the weight and size allowed on the LEO. The broad beam design may not require the beacon receiver FOV to be scanned, while the scanned beam will. For example, at values of P_{ave} = 50 mW and D_R = 15 cm, the maximum allowable θ_R is greater than the LEO attitude uncertainty for the broadbeam design, and ~1/4 the LEO attitude uncertainty for the scanned-beam method.

Thus, for the scanned-beam case, the beacon receiver FOV must be scanned over a 4 × 4 matrix to satisfactorily achieve acquisition. This increases the acquisition time by 16 × 2/N, assuming the FOV stays at each matrix element for at least two pulse periods to avoid missing the signal. For N = 10 pulses/sec, this increase is 3.2 sec, which is not unreasonable when compared to the 5 sec required to achieve steady-state tracking for N of 10 pulses/sec.

The advantage of the broadbeam method illustrates the requirement for high peak power and low-repetition-rate pulses for acquisition. For fixed average power, the peak power, P_k, is given by

$$P_k = P_{ave} \text{ / duty cycle} \qquad (3.14)$$

where duty cycle = the product of N $\Delta\tau$.

For the example considered here, the broadbeam method has a duty cycle of 10^{-7} and the scanned-beacon method has a duty cycle of $2.8 \cdot 10^{-6}$. A continuous wave (CW) laser has a duty cycle of 1, and a mode locked laser operating at 1.0 Gb/sec has a duty cycle of about 10^{-1}. These high duty cycles clearly indicate the reasons why the communication laser is not suitable for initial acquisition, and why the acquisition process should be started at the GEO.

While the broadbeam design may appear advantageous over the scanned-beam design, when implementing an actual project, the scanned case has a number of clear advantages. For example, the scanned case has a higher pulse rate, which is more desirable because it allows the tracking servo system to have a higher frequency response and consequently a smaller tracking error. Also, the higher pulse rate allows command data to be transmitted from the GEO to the LEO on the beacon beam. Thus, the laser used as the acquisition beacon in the scanned mode can also be used as the tracking beacon merely by stopping the scan and pointing the same beam at the LEO. The beacon receiver then receives all the N · M pulses per second. When the broad beam is narrowed, the beacon receiver picks up higher power pulses but at the same N pulses/sec rate.

There are also practical design components associated with scanning and the beam-broadening mechanisms, as well as their laboratory and field testing, to determine the final decision of which method to use. The current state of the art does, however, argue for the scanned case.

3.5 Step 2 of the Acquisition Process

After the acquisition beacon has been located by the LEO, the accuracy pointing of the communication laser toward the center of the FOV of the GEO's telescope antenna will depend on the servo loop's tracking accuracy. The relative orbital motions will result in the line of sight between the LEO and GEO having a maximum angular velocity of 1.5 mrad/sec and a maximum angular acceleration of 3.5 μrad/sec^2. A Type II servo is necessary to maintain an angular tracking accuracy under a constant acceleration. The tracking error

$$\theta_E = \ddot{\theta}_{max} / K\alpha \tag{3.15}$$

where

$K\alpha$ = gain constant of a Type II servo
$\ddot{\theta}_{max}$ = constant maximum acceleration, 3.5 μrad/sec^2

To point the 10-μrad beamwidth laser beam to 1/10 of its beamwidth would require a maximum tracking error, θ_E, of 1 μrad. For $\ddot{\theta}_{max}$ = 3.5 μrad/sec^2, the required $K\alpha$ would be 3.5/sec^2. This value of $K\alpha$ is about the best that can be obtained with a pulse rate of 10 pulses/sec while maintaining loop stability. Due to the fact that communication does not occur until the acquisition beacon has been replaced with a higher pulse-rate tracking beacon, it may be advantageous to initially broaden the communication beamwidth to allow for the fact that acceleration greater than 3.5 μrad/sec^2 may result from the attitude motions of the LEO itself.

However, as will be seen later on when we consider methods of extracting, by means of accelerometers, vibrational data and attitude data, due to assorted satellite movements, and their compensation, it will be found that we will not have to broaden the high-data-rate beam.

3.6 Step 3 of the ATP Process

After the LEO-based high-data-rate laser points its beam to the GEO, the tracking receiver of the GEO maintains its knowledge of where the beam emanation in the LEO is located. The signal-to-noise ratio (S/N) for the acquisition, S/N-acq, is compared with the S/N associated with the communications, S/N-com, measured at the GEO, and may be expressed as

$$\frac{(S/N)\text{-acq}}{(S/N)\text{-com}} = \frac{\theta\text{-com}(B\text{-com})^{1/2}}{\theta\text{-acq}(B\text{-acq})^{1/2}} \tag{3.16}$$

where

S/N-acq	=	S/N of the acquisition and tracking receiver on the LEO
S/N-com	=	S/N of the communication receiver on the GEO
θ-com	=	FOV of the data receiver on the GEO
θ-acq	=	FOV of the tracking receiver on the LEO
B-com	=	bandwidth of the high-data-rate communication receiver on the GEO
B-acq	=	bandwidth of the tracking receiver on the GEO

Optical communication systems analysis, based on the expressions developed in Chapter 2, has shown that the 1-Gb/sec data-rate signal can be obtained in the presence of sunlit Earth with a communication optical FOV greater than 100 μrad. If θ-acq = 7 milliradians, B-acq = 10 Hz, and B-com = 10^9 Hz, the S/N-acq is 140 times the S/N-com. Thus, even if the communication beam is spoiled during the acquisition phase, there will always be sufficient energy on the GEO to acquire it and track it.

The tracking function on the GEO serves to keep both the communication receiver FOV and tracking beacon pointed at the LEO satellite. The allowable tracking error is determined by the smallest of either one-third of the FOV-com or one-tenth of the tracking beacon beamwidth.

3.6.1   Use of the GPS to Determine the Location of the LEO

The position of the GEO satellite can be determined by its ephemeris from ground-station observations and orbital calculations, with updating as necessary, yielding accuracy of ~10s of kilometers, with an onboard star sensor providing attitude data. The position of the LEO can be obtained directly by employing the GPS satellites. LEO position accuracy of the order of a few meters is thus feasible, particularly when employing the differential GPS format. This makes pointing the beacon much easier, and achieves a more rapid ATP implementation.

Nevertheless, it is important to consider as backup, in the event of failure of the GPS receiver onboard the LEO, the use of ground stations to provide the ephemerides by ground stations and attitude data by means of SGLS receiver and onboard star sensor and inertial guidance instruments. In the advanced GPS design, it is anticipated that there will be upward-directed beams that will enable GEOs to determine their position with a degree of accuracy in the 10s of meters as well.

3.6.2 Acquisition Timing

Based on the previous discussion, the time required to complete each of the steps involved in the acquisition leading to the closing of the communication loop is given in Table 3.2. More time may be added to allow for switching operations and the possibility that the beacon FOV may have to be scanned under Moon background conditions. Even with this addition, it is possible to complete the acquisition process in less than 20 sec.

3.6.3 Acquisition Timing Using GPS

The use of the GPS onboard the LEO to provide its position to within approximately a few meters' accuracy, and having this data sent via RF (from

Table 3.2
The Acquisition Time Sequence

Function	Duration (sec)	Elapsed Time (sec)
1. Sequence starts at GEO	–	–
2. Beam travel to LEO	0.12	0.12
3. Acquisition of beacon on LEO (without scanning)	5.0	5.12
4. Beam travel to GEO	0.12	5.24
5. Acquistion of communication beam on GEO	0.5	5.74
6. Change beacon mode from acquisition to track	0.5	6.24
7. Beam travel to LEO	0.12	6.36
8. Check beam code, stop acquisition, and begin communication	1.0	7.36
9. Beam travel to GEO	0.12	7.48
Total time from start of beacon acquisition to receiving of communication stream: 7.48 sec		

the LEO to the GEO or from the LEO to a ground station and then retransmitted to the GEO) will enable the GEO's beacon to be directed at the LEO with the appropriate look-ahead angle compensation. It is estimated that this process can achieve the closing of the loop within less than 7 sec.

3.7 Additional Tracking Considerations

Tracking to accuracies in the range of 1 μrad may be difficult to achieve but is considered doable. As mentioned, a Type II servo loop will be required. It was also pointed out that the maximum obtainable Type II gain constant, $K\alpha$, is 3.5/sec^2 when the beacon pulse rate is 10 pps. If the tracking beacon pulse rate is increased by a factor of 100, it is possible to obtain a $K\alpha$ of 3.5 · 10^4/sec^2. Using (3.15) and value of $K\alpha$ = 3.5 · 10^2, a 1.0 μrad tracking error can be maintained under a constant acceleration of 35 mrad/sec^2. The accelerations resulting from the spacecraft's attitude control system, onboard vibrations, or motion of the laser communication system's coarse-pointing optical elements need to be below the value of 35 mrad/sec^2.

Additional discussion on the mechanical vibration and its isolation needs, as well as its effect on the ATP and communication link performance, will be presented in Section 3.11, the last section of this chapter.

In implementing tracking systems, consideration must also be given to the practical size of the available tracking photodiodes. Devices such as image dissectors, quadrant photomuliplier tubes, and quadrant silicon PIN diodes have sensitive areas about 2.54 cm in diameter. Silicon avalanche photodiodes have sensitive areas up to 0.127 cm in diameter, but multielement arrays are always possible. Without use of a fine-pointing mirror, the motion of a light spot focused on these detectors can be sensed accurately only to about 0.00254 cm, although some image dissectors and CCDs can achieve 0.00127 cm. The angular error, θ_E, and the detector linear position sensitivity, L_D, are related as

$$\theta_E = L_D / f' \tag{3.17}$$

where

L_D = linear position sensitivity of detector
f' = equivalent focal length of the receiving optical system as seen at the detector

A tracking accuracy of 1 μrad with a linear position sensitivity of $10^{0.3}$ inches would require an equivalent focal length of at least 2,540 cm.

However, as is well understood, the design of optical systems with such focal lengths can be constructed with package sizes much smaller than the equivalent focal length, by using microscope objectives as relay elements.

Although a large focal length is required for tracking, the same focal length cannot be used for acquisition unless the beacon receiver is scanned over the LEO satellite position and attitude uncertainty. If scanning is not used, the acquisition detector must view a field of at least 1° (~20 mrad). To cover this field with an equivalent focal length of 2,540 cm would require a 50.8-cm-diameter detector, which is, of course, impractical. Therefore, a smaller equivalent focal length (of 127 cm, for example) would be necessary for the acquisition detector.

The circuit configuration using separate acquisition and tracking detectors, as shown in Figure 3.8, will provide an acceptable design. It combines realistic and yet size-efficient optical and electronic components. The output of the acquisition detector is used to position the coarse-pointing mirror, while the output of the tracking detector controls the fine-tracking steering mirror positioner.

The optics for the tracking system need not be diffraction limited. It is only required that the image be circularly symmetric so that motion in any direction provides the proper error signals. Then a reasonable amount of spherical aberrations can be tolerated.

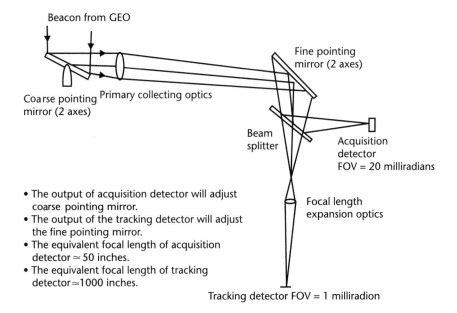

Figure 3.8 LEO acquisition and tracking system.

Aberrations such as coma and astigmatism, which cause image distortions that are not circularly symmetric, are not acceptable. However, when the system is in the tracking mode, the image is very near the optical axes. For the large F/number optical systems required, coma and astigmatism will be negligible over the small field angles being used.

3.8 Integration of the ATP Within the Laser Transceiver

Having discussed the basic functions of the ATP systems, specific integration schematics starting with the laser transceiver on the LEO are now described, followed by additional ATP integration with inclusion of inertial sensor systems to servo out much of the interfering vibrational spectra of the satellite platform and its equipment.

3.8.1 The Laser Transceiver for the LEO Satellite

The ATP for a laser transceiver on the LEO will be aimed by acquiring and tracking the beacon laser, which is located on the GEO. Referring to Figure 3.9, the LEO's laser transceiver has three ATP functions that need to be continuously performed:

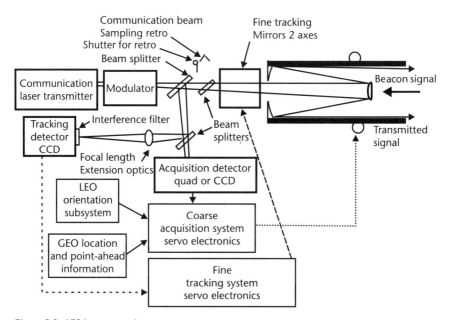

Figure 3.9 LEO laser transceiver

1. Coarse pointing
2. Fine pointing
3. Alignment between the GEO tracking detector and the laser transmitter beam

Initially, the LEO's acquisition subsystem and knowledge of the LEO's attitude information, the GEO's position, and the PAA information are used to adjust the gimbals of the primary telescope (shown as diffraction-limited Cassegrain in Figure 3.9) to receive the beacon signal from the GEO.

As emphasized in Chapter 2, the gimbals will require vibration isolation to prevent motions of the antenna or other large optical systems that might be on the LEO from affecting the laser communications systems' pointing accuracy. The gimbals use optical shaft encoders with absolute accuracies in the submicroradian range. Present day interferometric devices can easily achieve the required accuracies. A beam-splitter in the primary optical train serves to direct the beacon signal to the acquisition detector, which can be a quadrant photo detector or a CCD. If the beacon laser is at a different wavelength from the communication laser, the beamsplitter can be a dichroic mirror that passes the communication laser wavelength and reflects the beacon laser wavelength.

If the communication laser and the beacon laser are of different wavelengths, reflective optics are dictated for the primary telescope. If the two lasers operate at the same wavelength, a refractive primary telescope is possible. The same is true for the optics of the tracking detector's focal length.

Once the beacon is received by the acquisition detector, the differential location signals are sent to the acquisition subsystem signal-processing electronics. The signal-processing electronics send the servo signals to the telescope gimbals to keep the primary optical axis pointed in the direction of the incoming beacon laser. Fine pointing now begins.

Fine pointing is accomplished by a configuration of small mirrors located in the optical train so that the optical gain of the system increases the angular precision of each mirror. These small mirrors have precise angular motion, 90° from each other, so that two-axis alignment of the outgoing laser communication beam and the incoming beacon laser can be maintained within microradian accuracy in the presence of LEO satellite vibrations. When the fine-pointing mirrors get close to ends of their angular travel, the servo system moves the coarse-pointing gimbals so that the fine-pointing mirrors stay within range of the angular motion.

Precise alignment between the communication laser beam and the beacon laser tracking detector is critical to the operation of the laser communications system. This alignment can be facilitated by mounting the laser, modulator, and beacon laser tracking detectors on the same optical bench, which should be mechanically well isolated from the rest of the LEO satellite structure. The angular alignment between the communication laser outgoing beam and the beacon laser tracking detector should be checked periodically to ensure that the communication beam will be directed accurately enough to illuminate the GEO's communication receiver optics. (This periodic check should be done during a time when laser communications are not taking place so that the proper functioning of the beacon tracking system is not disturbed.) This alignment check can be accomplished by using a retroreflector located on the optical bench to reflect the communication laser output back onto the beacon tracking detector. Any misalignment can be corrected for by either noting the location of the communication laser return on the tracking detector CCD and updating the fine-tracking mirror servo system to make that location the point to which the beacon laser is pointing or by noting the offsets in the fine-tracking mirrors needed to position the communication laser to the point on the tracking detector CCD to which the beacon laser is pointed and making the corresponding offsets in the fine-tracking servo system. (The location of the retroreflector must be such that it can be blocked from reflecting the communication lasers back onto the beacon detectors when the actual communication is taking place.)

Based on Figure 3.9, it may be further seen that there are two major servo loops. One being the coarse-acquisition (10s of Hz) and the other, the FSM, which accommodates several 1,000 Hz. The two servo loops work together; the coarse handling large angular excursions and the fine, the small-amplitude, higher frequency excursions, to bring and keep the direction of the incoming beacon colinear with the LEO's primary optical axis. When this stage is reached, the high-data-rate communication laser on the LEO begins to transmit its high-data-rate signal beam to the GEO satellite.

3.8.2 Baseline Laser Transceiver with Inertial Sensors

Another manifestation of the ATP design (discussed by Lee, Ortiz, and colleagues [2, 3]) and shown in Figure 3.10, which is more the current design, considers the placement of inertial sensors on the satellite to detect the vibrations. After their processing, their outputs are used to develop an error signal to enable compensatory servo loops to take out the effect of vibrations on the laser beam.

As is continually emphasized in this book, these vibrations are introduced by the various mechanical equipment on board the satellite platform, such as the reaction wheels and solar arrays. Another source of vibrations, although of smaller frequency through higher amplitude, is the "dead band"

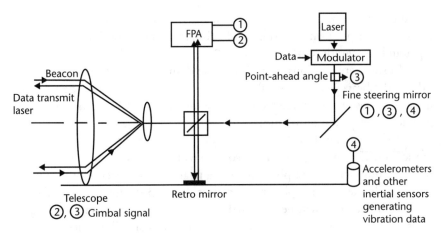

The difference between the estimated position of the beacon and that of the transmitter laser with a point-ahead angle becomes a pointing command to the Gimbal.

Figure 3.10 Laser ATP transceiver design employing accelerometers.

vibration of the platform that is introduced during its orbital motion. The dead band cycle of the spacecraft is roughly a pseudoharmonic motion of the satellite, as it weaves back and forth along its sides, while orbiting forward along its trajectory.

While the analysis and description in this book are given for spacecraft that are of medium weight (typically hundreds of pounds), the advent of microsatellites weighing on the order of 20 pounds would require micro and submicro components (often under the rubric of nanotechnology) to achieve the stability and isolation requirements in order to maintain the "sipping straw" laser communication between the spatial microsatellites. However, the principles and the equations discussed in this chapter will hold for small-size and small-weight platforms as well. But because the microsatellite field is just beginning its initial testing and the large-size spacecraft and equipment are here and have been on hand for many years, we will continue with a technical discussion of the ATP that will be consistent with the design of satellites that are typically at least a couple of hundreds pound in weight.

To continue, of the internal measurement unit (IMU) family of sensors, the primary inertial sensors that are used are the accelerometers. Their output is doubly integrated, thus providing a position versus time signal that will be subtracted from a singular reference signal to form the error signal. The error signals are used to produce adjustments of both the coarse-steering positioner and the fine-steering mirror positioner. It should be pointed out that JPL has found that it is a more accurate technique to do numerical integration of the accelerator output using the Trapezoidal Rule, rather than performing the analog double integration [4].

In some applications, it may be more desirable to let the Earth be considered the beacon; and in some cases from the Earth, it may be desirable to use a high power uplink laser system (assume one at 0.53 μm for this example) deployed on a mountaintop.

The telescope shown in Figure 3.11 will receive the Earth's reflectance of the Sun. The coarse-pointing servo loop will adjust the telescope to receive the maximum amount of photons from a relatively low-level photon source.

While the beacon splitter will collect the 0.53-μ signal, the balance of the earth's "beacon" will be collected onto the focal plane array. The signal there will be in the wavelength range of from 0.4–0.9μ. (The selected wavelength would depend on the solar spectrum reflected off the Earth and the Earth's atmosphere and the relative spectral transmission of the LEO's transceiver optics at wavelengths other than 0.5μ). Clearly one may select 1030

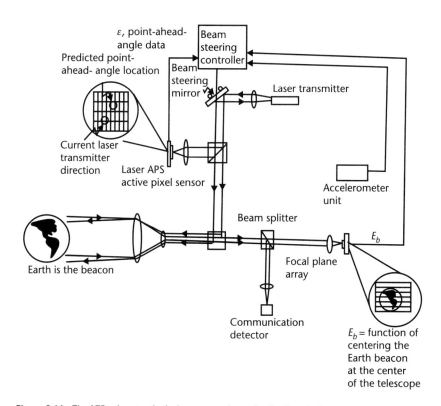

Figure 3.11 The ATP subsystem in the laser transceiver using Earth as the beacon.

µm and 1064 µm, or 1500 µm and 1550 µm, which may be obtained by using an Nd-doped YAG laser, various laser diodes, or fiber lasers [5].

The external jitter measured by the processed output signal of the accelerometer system will provide the vibration signal that is used to compensate for the spacecraft vibrations. And when the high-data-rate laser signal is transmitting, the fine-steering mirror will be adjusted (in a fast-tracking mode of up to several thousand times per second), with a servo system inputted with an error signal that is the difference between the PAA and the extracted vibration signal.

Another error signal is in the servo loop associated with the location of the transmitter laser and the predicted PAA. In Figure 3.11, both the slow and the fast loops produce "concentric" or simultaneous changes in the ATP to enable the adjustments necessary for maximizing the beacon signal reception and also the transmit laser output from the same telescope antenna.

In another configuration, two separate FPAs are used. A fraction of the transmit laser is used as the reference signal on the Transmit-FPA, and the beacon produces a position curve on the Receive-FPA. The resulting correction signals are used to cancel the vibration signal derived from the accelerometer subsystem. Together with the predicted PAA signal, they will be used to make adjustments in the fine-steering mirror and the coarse-pointing gimbals. The coarse- and fine-adjustments subsystems are slaved to one another.

The fast-tracking signal on the FSM will also be connected to the spacecraft's attitude control system (AC) to adjust its attitude direction.

Finally on the matter of mitigating the number of "noise" photons from the Sun entering the receiver optical system, a band interference filter centered at the communication laser wavelength is employed and placed over the CCD. There is one shown in front of the tracking detector in Figure 3.9. That filter is centered at the beacon wavelength to keep any scattered light from the Sun or the Moon that might get into the optical system from reaching the tracking detector. One could add this type of filter in front of the communication detector in the later figures that depict Earth as a beacon [6].

3.9 Summary and Concluding Remarks

Chapter 3 describes the ATP functions of the LEO satellite and the GEO satellite, starting from the PAA, the generation of a beacon signal and its search of the LEO. That is followed by the acquisition of the beacon by LEO and the transmission of a high-data-rate signal to the GEO by the LEO.

Attention is paid to the filtering of the Sun and Moon spectral radiance in the LEO to GEO link, the trades between the broadened beacon beam versus the scanned narrow laser beam, and estimate of the total time required from the turning on of the beacon to the transmission of the high-data-rate signal to the GEO.

Chapter 3 shows how the ATP functions may be integrated with the laser transceiver. This includes two major servo circuits: the coarse-acquisition (tens of Hz) and the fine-steering (thousands of Hz) mirrors. Several schematics are provided to show how the key subsystems are combined. Also shown is the use of accelerometers to detect the platform's vibrations and thence provide inputs to the servos, which result in minimization of the laser beam's jitter.

While the emphasis in this chapter is on the implementation of ATP functions in the LEO and GEO, the complexity of design being resolved in this application will find use in many other applications, for example, in the ATP design between GEO (or LEO) and airborne platforms such as fixed-wing or rotary-wing aircraft, between an LEO (or GEO) and airship or UAV, and also between LEO (or GEO, or aircraft) and ground-based fixed and moving stations.

References

[1]   Kraemer, Arthur R. "Acquisition and Angle Tracking of Laser Communication Links," McDonnel Douglas Memo prepared for the USAF Program 405B.

[2]   Lee, S., and G. G. Ortiz, "Atmospheric Tolerant Acquisition Tracking and Pointing Subsystem," *SPIE*, Vol. 4975, 2003.

[3]   Lee, S. J., W. Alexander, and G. G. Ortiz, "Sub-Micron Pointing System Design for Deep Space Optical Communication," *SPIE*, Vol. 4272, 2001.

[4]   Arnon, S., S. R. Rotman, and N. S. Kopeika, "Performance Limitations of a Free Space Communication Satellite Network Owing to Vibrations: Heterodyne Detection," *Applied Optics*, Vol. 37, No. 27, September 1998.

[5]   Guelman, A., A. Kogan, A. Kazarian, A. Livne, M. Orenstein, H. Milchalik, and S. Arnon, "Acquisition and Pointing Control for Inter-Satellite Laser Communications," IEEE Transaction on Aerospace and Electronic Systems, Vol. 40, No. 4, 2004.

[6]   Russell D., H. Ansari, and C. C. Chen, "Laser Pointing, Acquisition and Tracking Control Using a CCD-based Tracker," in *Free Space Laser Communication Technologies VI, Proc. of SPIE*, Vol. 2123, 1994, pp. 294–303.

4

Satellite Downlink Through the Atmosphere

4.1 Introduction

Chapter 4 examines the effect on the downlink communication link, going from a spacecraft to an earth-based optical station. The uplink, from a ground station to a satellite, is discussed Chapter 5, while the terrestrial links through the atmosphere, including the effect of weather on the signal loss, are discussed in Chapter 6.

Following this introductory section, the downlink analysis, composed of the evaluation of the satellite-to-ground station link, is presented in Section 4.2. It is shown that the downlink beam spreads geometrically, primarily explained by the laser optical beamwidth times the distance to the Earth, with only a small spread due to the atmospheric scatter and the variation in the beam steering. In the analysis, it is shown that due to atmospheric turbulence and beam jitter, the BER on the downlink for OOK modulation may be explicitly expressed. We show that the effect of turbulence is generally small on the downlink, and when no vibration is superimposed on the radial component of the downlink beam, the BER becomes even smaller. But when the vibration field occurs, it may be zeroed out. The latter was discussed in Chapters 2 and 3.

In our analysis, we specifically address an evaluation of the BER versus the RMS of the intensity of the atmospheric turbulence combined with laser beam's jitter. This is done in order to provide consistency of discussion of the atmospheric turbulence involved in the downlink connection from the spacecraft to the ground station or atmospheric-borne platforms (Figure 4.1). In the latter case, for airplanes and UAVs, depending on their speed and

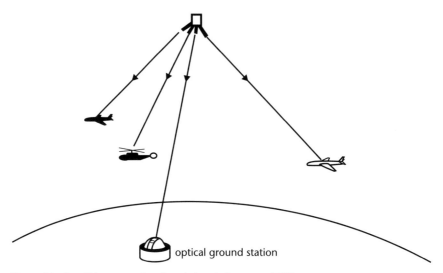

Figure 4.1 Downlinks to ground station, airplane, helicopter, and UAV.

altitude, we also get additional signal loss and consequent increase in BER, because of the layers of charged particles that often cover those aircraft.

Apart from providing downlink information to the ground and airborne vehicles, the downlink signal can also be used as a "reference" signal to the adaptive optical system (AOS), and thereby provide a way of aiding the needed adjustments that are required for the AOS of the ground and airborne optical terminals. Shown in Chapter 5 are the adjustments and movements of optical elements to enable a much better performance of the uplink laser signal, which is beamed from the ground station to the satellite.

The downlink signal suffers very small losses as its beamwidth spreads from the attainable diffraction-limited satellite's optics and goes through essentially a nonatmospheric path, until it reaches about 30 km from the Earth. By comparison, the losses of the uplink are very large because the beam begins to spread and accumulate distortion the very instant the photons are emitted from the ground-based telescope aperture. That is, as soon as the beam interacts with the atmosphere, the beamwidth broadens with increasing height, together with scintillations and beam wander, as the beam travels upward through the atmosphere.

In the discussion that follows, techniques are described that ameliorate the signal loss and the distortion problems. They involve the measurement of the effect of the atmospheric turbulence on the beam and then employ adaptive optics (AO) to perform the necessary corrections. AO uses either a combination of the downlink of the communication signal itself as a "reference" source or other reference sources, such as a star or an artificial laser reference. The latter may be the sodium layer of the upper atmosphere, which is excited at an altitude of ~90 km, by means of a ground-based laser directing energetic pulses into that atmospheric region.

We typically correct the uplink distortion by using the Reciprocity Theorem, which permits us first to measure, then to correct the downlink distortion by maximizing the downlink signal by means of adjustments of the deformable mirror of the AO subsystem. Then the uplink signal is transmitted through the same optics, thereby superimposing a distortion on the emitted uplink signal, which on its way to the satellite is combined with the distortion of the atmosphere occurring along the same path (through which the downlink beam has just gone through). The uplink signal will thus have its distortion roughly canceled, resulting in a near-plane wave when reaching the satellite aperture. That is, with the AOS, the uplink signal can be received at the satellite with relatively little distortion.

4.2 Downlink from Satellite to Ground Station

The laser beam going from a geosynchronous satellite to a ground station, shown in Figure 4.2, diverges through an assumed 10× diffraction-limited

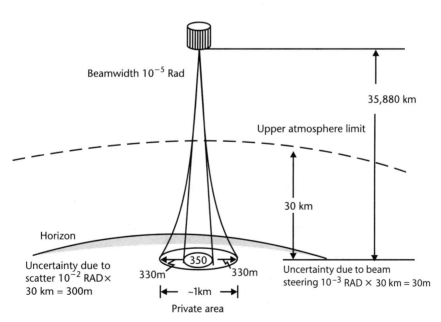

Figure 4.2 Satellite-to-ground link, comprising a long nonatmospheric passage and a small atmospheric link.

optics to provide a small circular area on top of atmosphere, at 30-km height above sea level. Below this height and down to sea level, special atmosphere-dependent losses, including scatter and tilt, begin to distort the downlink photon stream. In fact, with a 1.06-μ-signal wavelength, and from an assumed 1-m aperture at the satellite, we get for the synchronous distance of about 35,880 km a theoretical circle of ~350m diameter, consisting of uniformly radiating photons on the ground. Plus, more than 300m are added to that 350m diameter due to the effect of scatter, with another 30m due to the variation in the angle of arrival. The total spread of the beam intercepting the ground at the equator will be of the order of 1 km.

In the geometry of this example, the ground-based station is directly under the satellite. As we will see later on, a beacon is transmitted down from a satellite platform, which provides a reference to the optical ground station (OGS). Upon adjustment of the reference phase and other distortion components such as changes in angle-of-arrival and scatter, the AO telescope antenna system (Figure 4.3) delivers the best attainable signal level, with distortion correction. This would result in a lower bit error rate for the beacon's data stream. Additionally, by means of a heuristically accepted "law of reciprocity," if an uplink laser signal is transmitted through the same AO optics that have just been adjusted and corrected for the downlink, the uplink photons will follow in the same path that has been used and corrected for by the downlink photons' signal. That is, a combination of a modified-shape plane wave of the uplink going along the same path in the atmosphere that dis-

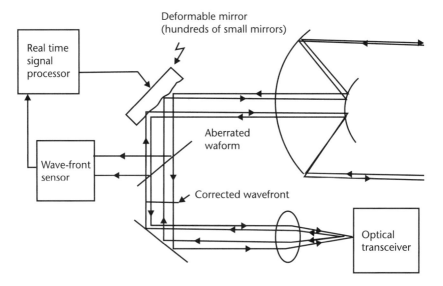

Figure 4.3 Principal components of the adaptive optics system.

torted the downlink will result in the beam coming out of the atmosphere and on to the satellite, which has been corrected by the AOS and combined with the atmospheric distortion. The beam will achieve a minimum of deviation from plane wave performance.

There are several methods of bounding the minimum distortion on the downlink and uplink using the previously described reference-adaptive optics and reciprocity (RAOR) system approach. The downlink reference may be a beacon located on the same satellite platform or on a separate satellite platform or any of the substitute references such as a star, or an artificial "star" reference source at ~90 km altitude—this might be a sodium laser activated by a ground-based laser.

Another reference design is composed of a space-based reflector mirror (see Chapter 8) reflecting a laser reference beam, derived from a laterally located satellite, to the ground station containing the AO subsystem. However, any of the selected references needs to be geometrically close enough to the downlink data stream directed to the optical ground station. The separation distance between the two platforms (supporting the reference and the signal downlinks) must be within a coherent angle measure, that is, within the isoplanatic angle.

4.3 Analytic Expressions of the Downlink Signal

In this section, the effect of the strength of turbulence on the downlink signal, with OOK modulation, is evaluated and the BER is calculated. From a physical description of how a beam goes down through the ~ 30 km of atmosphere, we consider the change in its direction from a downward straight line as it moves and changes direction. This occurs because it has to penetrate, diffract, and refract when interacting with the various molecular globules of atmosphere, each of a different index of refraction (varying in size, density, temperature, and humidity and also spatially and temporally). The beam therefore spreads in width and scintillates, as well as changing its angle of arrival at the receiver optics.

The term C_N^2, which is typically used throughout all discussions of laser beam propagation through atmospheric turbulence, may be best considered as a quantity proportional to the variance of the index of refraction fluctuation. This variance is taken between two points in the fluctuating atmospheric medium. It is often expressed as [1]

$$C_N^2 = \left[79 \cdot 10^{-6} p / T^2 \right]^2 C_T^2$$

where

C_T is the temperature structure constant

$$C_T = \left(\left\langle \left(T_1 - T_2\right)^2\right\rangle\right)^{1/2} r^{-1/3}$$

T, T_1, T_2 = the temperatures in kelvin. T is the running variable between T_1 and T_2

p = pressure in millibars

C_N = refractive index structure "constant," in $m^{-2/3}$

r = distance in centimeters in between which two measurements of temperatures, T_1 and T_2, are made

The best way to express the index of refraction between two points is to assume it to be the sum of an average value of the index, n_0, and stochastic component, n_s [2]. Thus,

$$n_T(r) = n_0 + n_s(r) \tag{4.1}$$

where [1]:

$$n_0 \approx 1 + 77p/T\left[1 + 7.53 \cdot 10^{-3}/\lambda^2 - 7733q'/T\right] \cdot 10^{-6} \tag{4.2}$$

r = location in space

p = air pressure in millibars

q′ = humidity in grams per cubic meter

T = temperature in degrees Kelvin

λ = wavelength ion meters

The refractive index of the spatial correlation is defined as

$$\Gamma_N\left(r_1, r_2\right) = E\left[n\left(r_1\right), n\left(r_2\right)\right] \tag{4.3}$$

where the symbol E is the average value. That is, we take the average value of two vectors.

Further by taking the Fourier transform of (4.3) we obtain

$$\Phi(K) = 0.033C_N^2 K^{-11/3} \tag{4.4}$$

where C_N^2 may further be described as the index of refraction structure constant and is a measure of the strength of the fluctuations of the refractive index. It is typically a function of the temperature between two points. It is also a function of pressure between those two points, but the pressure value is commonly constant, whereas the temperature does vary and therefore results in the beam moving in one direction and then another. Also, an examination of the measured value of C_N^2 indicates an inner scale and an outer scale at different times of the day. It is expressed in units of meters, raised to the $-2/3$ power. Weak turbulence is taken to be a value of $\sim 10^{-17}$ and strong turbulence, a value of $\sim 10^{-13}$.

K = vector wave number, which represents spatial frequencies

Additionally, the C_N^2 for height-related components may be written as shown [3]:

$$C_N^2(h) = A\exp(-h/H_A) + B\exp(-h/H_B) + Ch^{10}(h/H_C)$$
$$+ D\exp\{-(h-H_D)^2/2d_C^2\}$$

(4.5)

where

A = coefficient for the surface boundary layer and H_A = the height of its $1/e$ decay

B = coefficient for turbulence in the troposphere at up to ~ 10 km and H_B = the height of its $1/e$ decay

C = coefficient for turbulence at the tropopause and H_C = height of its $1/e$ decay

D = coefficient of turbulence of one isolated layer of turbulence, $(h-H_D)$ is the height of its $1/e$ decay, and d_C is its thickness

The covariance, for a plane wave for the case of the downlink laser beam from a satellite to a ground station over a path from L = 0 to L = Z with x as the running variable, may be expressed as [4]

$$\sigma_\chi^2(Z) = 0.56(2\pi/\lambda)^{7/6}\int_0^Z C_N^2(x)(Z-x)^{5/16}\,dx$$

(4.6)

The density distribution function of χ is normal and is then expressed as

$$f_\chi(\chi) = \left(1/\left\{\sqrt{(2\pi)}(\sigma_\chi)\right\}\right)\exp\left(-\{\chi - E[\chi]\}^2/2\sigma_\chi^2\right)$$

(4.7)

where the normalized received power is related to the log amplitude χ, as

$$I = P/P_o = \exp\{2\chi - 2E[\chi]\}$$

(4.8)

Equation (4.7) can now be added as a factor to the SPB expression for the downlink. From there we can proceed to the calculation of the BER [5].

In a considered example, the satellite downlinks a signal to an Earth station. The satellite is assumed to be in a circular orbit of 800-km radius, with a data rate of 1 GB and a wavelength of 1.55μ having a power output of 2W with OOK modulation. Additional parameters in the example include receiver aperture diameter of 1.2m, transmitter optics beamwidth of 500 μrad, and detector sensitivity of –47 dBm.

The general expression for the BER can be stated as [6]

$$\text{BER} = \int\int\{P(\text{on})P(\text{off}\,|\,\text{on},\theta,I) + P(\text{off})P(\text{on}\,|\,\text{off},\theta,I)\} \quad (4.9)$$
$$f_\theta(\theta)f_\chi(\chi)d\theta d\chi$$

where

P{off/on, θ, I)} when "1," that is, an ON signal is transmitted
P{on/(off, θ, I)} when "0," that is, OFF, or no transmission, takes place

For further simplification, we add the following notations [7]:

$$\alpha' = \left[P_T G_T R_{PD} / \{2\sqrt{2}\cdot\sigma_N\}\right]\cdot\left[L_T L_R (\lambda/4\pi Z)^2 G_R L_A\right] \quad (4.10)$$

L_A = loss in the atmosphere
R_{PD} = detector responsivity

We further include the substitution of

$$u = \theta^2/2\sigma^2$$

where

θ = the radial pointing error angle, without bias
σ^2 = the variance of the radial pointing angle, without bias.

We also substitute for v:

$$v = \chi - E[\chi]/\sqrt{2}\cdot\sigma_\chi$$

where

χ = the real part of the logarithm of the perturbation exponent as the log amplitude fluctuation

σ_χ^2 = the covariance for the plane wave coming through atmospheric perturbation from the satellite to ground over a path of length Z

The complementary error function may be represented by

$$\text{erfc}(a) = 2/\sqrt{\pi} \int_0^\infty \exp(-b^2) db \qquad (4.11)$$

By making all these substitutions, the equation for BER becomes [8, 9]

$$\text{BER} = 1/2\sqrt{\pi} \int_{-\infty}^\infty \int_{-\infty}^\infty \text{erfc}\left\{ \alpha \cdot \exp(-2 \cdot u \cdot \text{Gt}\sigma^2 + \sqrt{2}\sigma_\chi v) \right\} \qquad (4.12)$$
$$\cdot \exp(-u - v) du dv$$

where the first integral has limits from $-\infty$ to ∞ and the second integral has limits from 0 to ∞.

The BER is, in fact, BER(σ_χ, $G_t\sigma_\theta^2$). The plot of the BER versus the turbulence parameter, σ_χ, with the vibration of $G_t\sigma_\theta^2$ as parameter, is shown in Figure 4.4. As seen, for the case of zero jitter and turbulence of $\sigma_\chi = 0$, the

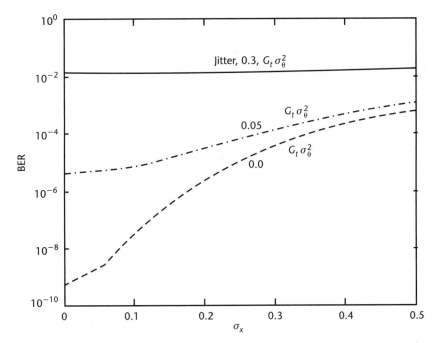

Figure 4.4 BER versus RMS of turbulence strength with jitter (vibration of beam along the radial axis from the satellite to the Earth terminal) as parameter $G_t\sigma_\theta^2$ from 0 to 0.3.

BER ~2 · 10⁻¹⁰. As discussed in Chapter 3, there are ways of reducing the effect of vibration of the radial pointing; therefore, very low BER may be feasible. However, in the plot of BER versus the RMS of turbulence strength with jitter (vibration of beam along the radial axis from the satellite to the Earth terminal) corresponding to parameter $G_T\sigma_\theta^2$, its variation, when taken from 0 to 0.3, will produce the results shown in Figure 4.4.

For a high level of vibration, $G_T\sigma_\theta^2 = 0.3$, the BER is a large constant (10^{-2}) even when the turbulence measure, σ_χ is anywhere between 0 and 0.5. However, as the vibration begins to diminish, for example $G_T\sigma_\theta^2 = 0.05$, the BER goes down to less than 10^{-5} at turbulence of 0.1, and down to less than 10^{-9} with the lower turbulence. However, as the RMS of the turbulence goes from 0.2 to 0.4, together with a vibration of 0.05, the BER increases from 10^{-5} to 10^{-3}.

Another aspect of the distortion of the downlink signal is expressed by the fluctuation of the angle of arrival, which occurs at the skirts of the downlink beam covering the ground station. This effect is discussed next.

4.4 Variation of Angle of Arrival of the Downlink Signal [8]

The fluctuation of the angle of arrival of the downlink signal aimed at the ground station or the airborne vehicle is a measure of the direction of the bulk of the energy of the photons relative to the plane of the aperture of the receiver. The fluctuation of the angle of arrival may be written in terms of the phase structure function. For example, let ΔS denote the total phase shift across the collector lens of diameter D, and let ΔL be the corresponding optical path difference.

Thus we may write:

$$k\Delta L = \Delta S \tag{4.13}$$

Now for small β, $\sin\beta = \beta$, so that as seen from Figure 4.5, we have [1, 2]

$$\beta = \Delta L / D = \Delta\beta / kD = \Delta\beta\lambda / 2\pi \cdot D \tag{4.14}$$

And further, for $<\beta> = 0$, we can assume that the variance of the angle of arrival would be

$$\left\langle \beta^2 \right\rangle = \left\langle (\Delta S)^2 \right\rangle / (kD)^2 = D_S(D,L) / (kD)^2 \tag{4.15}$$

where $D_S(D,L)$ is the phase structure function.

For the case of a plane wave and the Kolmogorov Spectrum, we can write the following two equations:

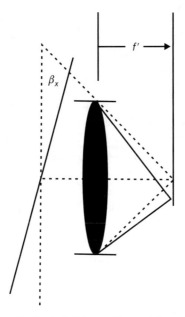

Figure 4.5 Illustration of the angle of arrival of the wavefront and the resulting dancing of the image.

$$\left\langle \beta^2 \right\rangle = 1.64 C_N^2 L l_0^{-1/3} \text{ for } D << l_0 \qquad (4.16)$$

$$\left\langle \beta^2 \right\rangle = 2.91 C_N^2 L D^{-1/3} \text{ for } D >> l_0 \qquad (4.17)$$

It should be noted that (4.16) and (4.17) are independent of wavelength and are given for the case of weak turbulence, but in fact hold for strong turbulence as well.

In terms of image dancing, it should be pointed out that previous expressions would also be applied, except that it would take place in the focal plane of the sensor. That is, the RMS image displacement would be the RMS of the angle of arrival $\sqrt{<\beta^2>}$ multiplied by f′, the focal length.

4.5 Summary and Concluding Remarks

Laser beam propagation through the atmosphere is shown to have special properties, depending on whether the signal is going downward, from a space platform to Earth, or upward, from Earth to the space platform.

The downlink, by virtue of its broadening when going through near-vacuum for a very long distance before going through the atmosphere, has little beam spread, beyond the geometric spread of $(\lambda/D) \cdot$ (distance). Therefore, the BER following its reception by the ground-based optical station is low. Nevertheless, using an adaptive optical system would improve the downlink signal reception, reduce the distortion, and, consequently, further diminish the BER.

In fact, we demonstrate this fact analytically; namely, that BER for low levels of RMS of the strength of turbulence ($\sigma_R \leq 0.1$) and low levels of beam jitter ($G_T \sigma_\theta^2 \leq 0.05$) extends from ~$10^{-6}$ to ~$2 \cdot 10^{-9}$ errors per bit. The vibration of the beam and resulting jitter can be greatly reduced by zeroing them out by means of the initial sensors and servo circuits discussed in Chapter 3.

At the end of the chapter, an expression for the fluctuation of the angle of arrival of the photon signal is derived. However, with an adaptive optics subsystem in the ground station, it is seen that the overall performance of the downlink signal reception would be enhanced.

The downlink signal described in this chapter did not consider weather interference or any man-made interference. However, it is clear that "data dumping" from an orbital platform can be transmitted to various locations on the Earth if they are weather-free. In Chapter 6, the weather avoidance system (WAS) is described. The chapter will show that WAS can help to achieve high performance of the downlink provided an integrated fiber cable is maintained between the various optical ground stations. This form of diversity reception, together with near-real-time weather information transmitted to the satellite platform, will significantly enhance the quality of the data transmission.

References

[1]   Toyoda, M., M. Toyoshima, T. Fukazawa, T. Takahashi, M. Shikatani, Y. Arimoto, and K. Araki, "Measurement of laser link Scintillation between ETS-VI and Ground Optical Station," *Free Space Laser Communication Technologies IX, Proc. SPIE 2990,* 1997, 287–295.

[2]   Kopeika, N. S., *A System Engineering Approach to Imaging,* Bellingham, WA: SPIE Press, 1998.

[3]   Andrews, L., and R. Phillips, "Laser Beam Propagation Through Random Media," *SPIE,* 1998.

[4]   Reyes, M., S. Checa, A. Alonso, T. Viera, and Z. Sodnik, "Analysis of the Preliminary Optical Links Between Artemis and the Optical Ground Station," *SPIE Proc.,* Vol. 4821, 2002, pp 33–43.

[5]   Toyoshima, M., and K. Araki, "The Far Field Pattern Measurement of an Onboard Laser Transmitter by Uuse of Space to Ground Optical Link," *Applied Optics,* Vol. 37, No. 10, 1998, pp 1720–1730.

[6]   Arnon, S., N. S. Kopeika, D. Kedar, A. Zilberman, D. Arbel, A. Livne, M. Gudman, M. Orenstein, H. Michalik, and A. Ginati, "Performance Limitation of Laser Communication Due to Vibration and Atmospheric Turbulence: Downlink Scenario," *International Journal of Satellite Communications and Networking,* Sept. 2003.

[7]   Kiasaleh, K., "On the Probability Density Function in Free Space Optical Communication Systems Impaired by Pointing Jitter and Turbulence," *Optical Engineering,* Vol. 33, No. 11, 1994, pp. 3748–3757.)

[8]   Sasiela, R. J., *Electro-Magnetic Wave Propagation in Turbulence,* New York: Springer Verlag, 1994.

[9]   Tyson, R. K., *Principles of Adaptive Optics,* Philadelphia: Academic Press, 1991.

5

Uplink Laser Communication Through the Atmosphere

5.1 Introduction

Chapter 5 starts with a restatement of the difference between the down-link and the uplink and leads to the expression for the criteria of when the AO system is needed in order to reduce the distortion of the uplink. As we will see, it depends on the ratio of aperture of the transmitter, D, to the lateral coherence length, ρ_o, in the aperture plane: If the coherence length is much larger than the aperture diameter, then we can do without the AO system. But when $D > \rho_o$, we need the AO system. However, as we will see, the AOS is generally beneficial and can do us no harm, whether in the optical receiver design of the downlink, uplink, or over terrestrial links.

Following C_N^2, the second major term in the analysis of the propagation of a laser communication beam through turbulent atmosphere, is Fried's coherent length. It is expressed by the symbol ρ_o or r_o. Either of those terms, is defined as the coherent length and may be expressed as

$$\rho_o = \left[0.423 k^2 \sec Z \int_o^L C_N^2 (h) dh \right]^{-3/5}$$

where

h	=	altitude
Z	=	zenith angle
L	=	path length

k $= 2\pi/\lambda$

C_N^2 = atmospheric structure constant

It is important to note that with the coherent length greater than the diameter of the receive optics, there will be a relatively small distortion in the receive signal. But with Fried's length smaller than the diameter, the distortion suffered by the signal could be severe.

The reference downlink in our AO systems may be one of several design configurations of artificial space based sources. Both primary and secondary sources are considered. Also included in this chapter will be measurement approaches for obtaining C_N^2. Finally, we present an example of how large the signal loss on the uplink can be if no AO are used, that is, when the ground station diameter exceeds the signal's coherence diameter.

5.2 Differences Between Downlink and Uplink

As we have seen from Figure 4.2, when going from a spatial platform to an earth station, through the atmosphere, there is a very small loss beyond the diffraction spread due to the optics and the range to the earth terminal, the scatter and the angle of arrival fluctuation. This is because the beam traverses some 40,000 km without engaging any atmosphere until we get to the last, roughly 30 km. However, the uplink beam starts out in the dense atmosphere and becomes distorted by the spatial and temporal changes in index of refraction, from the very start of the beam's emanation from the ground based telescope antenna.

Dr. Hal Yura, who has been the atmospheric turbulence guru at the RAND Corporation and also at the Aerospace Corporation and is a premiere mathematical physicist, describes a way of best remembering the difference between the laser downlink communication and the laser uplink communication.

The downlink is like a person entering a bathroom in which a shower curtain is drawn about an individual who is in the midst of taking a shower (Figure 5.1). The person entering the bathroom comes in with a flashlight turned on, and is aiming it at the shower curtain. The light will enable him to see (that's the downlink) the dark shape of the figure showering and all the movements of the limbs of the person luxuriating in the "waterfall." But the person in the shower cannot see (the uplink) past the running water flow, and thus cannot see who or what has just come into the bathroom. (Yura has also contributed a number of key mathematical relationships associated with electromagnetic signal penetration of turbulent atmosphere, which the present author has borrowed for this book.)

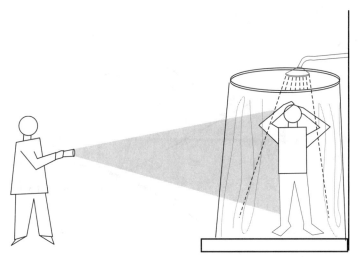

Figure 5.1 Conceptual analogy between uplink and downlink employing the "shower curtain" effect.

5.3 Calculating Signal Coupling Efficiency

For practical purposes, in order to obtain Γ'_T, the signal coupling efficiency factor, which needs to be inserted in the SPB equation, the ρ_o, the turbulence-induced lateral coherence length in the aperture, must first be evaluated. For this evaluation we need to express C_N^2, the atmospheric structure constant, as a function of the corresponding Z, the zenith angle.

The zenith angle is the angular spread from the perpendicular through the OGS to the plane running from the OGS through the spatial platform, or through an atmospheric platform with which we are to communicate. Those platforms may include an aircraft, UAV, helicopter, or airship (Figure 5.2).

Clearly, as the zenith angle may move from 0 degrees (where the satellite is perpendicular to the horizontal going through the OGS) to larger values by virtue of the platform's movement along its trajectory, the photons' path along the atmosphere and space increases. As is evident, at a zenith angle that is $\geq 85°$, the distance becomes exceedingly long, and so are the losses, making the considered link no longer practical. Figure 5.3 shows the plot of the ρ_o versus the zenith angle.

The ρ_o, the lateral coherence distance, is observed at the ground transmitter aperture due to a point source at the satellite. The values from Figure 5.3 can be used to evaluate the atmospheric coupling coefficient Γ'_T, for the uplink.

Shown in Figure 5.4 is a plot of Γ'_T as a function of the beam's zenith angle for aperture diameters of 14 cm and 25 cm. The expression for Γ'_T, the coupling

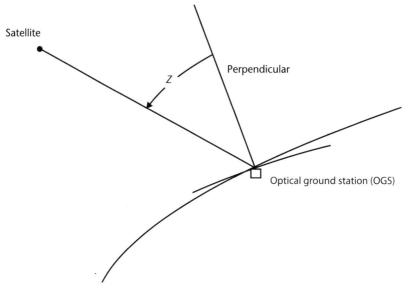

Figure 5.2 Geometrical description of Z, the zenith angle, that is required for the evaluation of ρ_0, the lateral coherence distance.

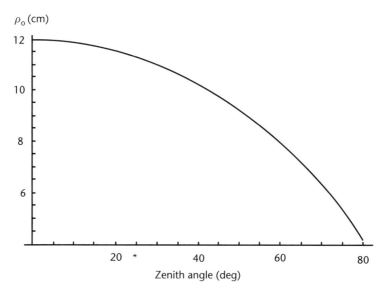

Figure 5.3 Plot of the lateral coherence distance versus zenith angle.

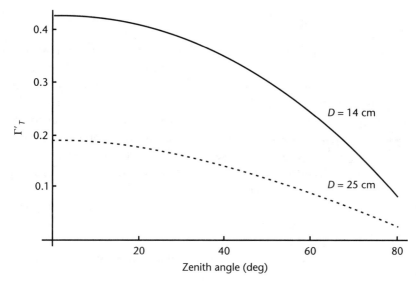

Figure 5.4 Plot of the turbulence induced signal coupling efficiency, Γ_T, versus the zenith angle. The Γ_T is a factor that should be inserted in the signal power budget for the evaluation of signal strength in the atmosphere [6].

efficiency of the uplink signal from the ground station to the satellite, may be expressed as

$$\Gamma'_T \sim 1/\left\{1+\left(D/\rho_o\right)^2\right\} \tag{5.1}$$

where

$\quad D \;=\;$ collector aperture at the satellite
$\quad \rho_o \;=\;$ coherence length

When the basic signal photon power budget (Chapter 2) is combined with the Γ'_T factor, the uplink signal power budget becomes:

$$n'=\left\{P_T G_T L_T G_R L_R L_P Q\right\}\Gamma'_T/\left\{Lsfh\nu\right\} \tag{5.2}$$

where n' = number of photoelectrons per bit.

Clearly, from (5.1), if the $\rho_o \ll D$, that is, if the lateral coherence length is much smaller than the aperture diameter, that will imply a lot of phase distortion and in consequence $\Gamma'_T \ll 1$. This would the dictate the need for an adaptive optical subsystem in our OGS. The AO system will correct for the phase front distortion. And with the additional function of aperture averaging, correction of the signal fluctuation due to turbulence induced scintillation will be achieved.

The adaptive optical system corrects the phase front distortion by making use of the principle of reciprocity, that is, the ground-based laser transmitter beam goes through the same optical antenna as the received downlink reference beam. The reference beam distortion is reduced or corrected by virtue of adjustments of the various tiny sections of the deformable mirror, so that when the uplink photon stream emerges from the adaptive optics telescope, it will go through the same path that the downlink signal just went and arrive at its satellite receiver with a minimum of distortion. In other words, the uplink signal will go out with a particular set of distortions that is equal and opposite to the distortions suffered by the downlink when going through the atmosphere. The uplink interaction with the atmosphere will thus cancel much of the distortion.

We also add to the laser terminals located in the atmosphere and at the ends of the Earth-to-satellite links an aperture-averaging subsystem (AAS), whose function is to reduce the intensity of fluctuations generated by turbulence-induced scintillations.

5.4 Coherence Length and Associated Atmospheric Turbulence

The expression for coherence length [1], for $Z \leq 85$ degrees, and the standard Hufnagle-Valley Profile, with $\lambda = 1.06\mu$, is given in (5.3):

$$\rho_o = \left[0.423 (2\pi/\lambda)^2 \sec Z \int_o^\infty C_N^2 (h) dh \right]^{-3/5} \qquad (5.3)$$

Shown in Figure 5.3 is an illustration of ρ_o versus the zenith angle, implying that when ρ_o is greater than D, it would reduce, and possibly eliminate, the need for an adaptive optical system. But at the same time, the diameter of the aperture has to be large enough to collect as many signal photons as possible. It is therefore necessary to get an intermediary size, for example, of the order of 10 to 20 cm, for the aperture.

From the various probabilistic models that have been developed, the one most commonly used today, which best describes the atmospheric turbulence, $C_N^2(h)$, is known as the Hufnagel, Valley, and Bufton Model, and is given by Sasiela [2] as

$$C_N^2 (h) = 0.00594 (W/27)^2 (10^{-5} \cdot h)^{10} \exp(-h/1000) + 2.7 \cdot 10^{-16} \\ \times \exp(-h/1500) + A \exp(-h/100) \qquad (5.4)$$

where W= the RMS of the wind velocity at altitude h; it is often set at 21 m/sec.

The turbulence strength is typically taken as zero at h > 30 km; however, it is most dependent on W. Typically the model's nomenclature is HV-21 and implies a wind velocity as 21 m/s. This model is often written as $H_{5/7}$, wherein the coherence diameter is ~5 cm and the isoplanatic angle is 7 μrad.

5.5 Measurement of Atmospheric Effects on Downlink and Uplink

The basic measurements of the Atmospheric Structure Constant, temperature gradient and wind velocity at altitude of up to one kilometer, were initially performed by Culman [3], using a tethered balloon. This data is coupled with subsequent data measurement at altitudes greater than 1.0 km, so that complete atmospheric profile of its structure constant can be gotten under a variety of meteorological conditions.

It is often desirable to add measurements using an aircraft platform in order to perform scintillation measurements as a function of the transmitted beam divergence. In particular, it is necessary that the scintillation be measured for 100 μrad down to the diffraction limited transmitter optics, which is typically ~3 μrad. As mentioned above, these measurements are to be performed under different meteorological conditions. Also, it should be noted that in the measurement field, with large beam angles the scintillations may be much smaller than with the smaller beam angles.

Additionally it is necessary that beam spread measurements be made. For the uplink, the beam spread measurements could be made by focusing the earth based laser beam at the measuring aircraft and then determining the resulting spot size. This can be done by sweeping the uplink beam by the receiver, or vice versa. The resulting spot size (S) is given by

$$S \cong \lambda F \left[1/D^2 + 1/\rho_o^2 \right]^{1/2} \tag{5.5}$$

where λ is the wavelength, F is the focal range which in our measurement case is roughly 20 km, D is the diameter of the output laser beam, and ρ_o is the 1/e of the atmospheric modulation transfer function (MTF) of a point source located at the aircraft and observed at the transmitter site, namely at the ground-based transmitter laser system. In the absence of turbulence (that is when $\rho_o \cong \infty$),

$$\text{spot size} \cong \lambda/D(F) \tag{5.6}$$

For example at $\lambda = 0.53 \,\mu$m, F = ~20 km, and D = ~20 cm, the spot size is found to be ~5 cm. However, in the presence of turbulence, the spot size is spread according to (5.5). Then by comparing the resulting spot size to its value in the absence of the turbulence, we infer the value of ρ_o. The latter characterizes the effect of turbulence on the beam spread, scintillations and beam wander.

For the downlink, a method that enables one to determine the MTF of the atmosphere for a very small modification of the experimental package is the following: The aircraft-based laser beam illuminates the receiver on the ground. The input lens of the receiver on the ground is assumed to have a focal length F. Then, it can be shown that the one-dimensional Fourier transform of the intensity I(x), as measured through a slit scanner in the image plane of the receiver, is given by the product of the MTF of the atmosphere and the MTF of the lens [4]. That is,

$$\left\{ M_{\text{atmosphere}} \left(FK/k \right) \right\} \left\{ M_{\text{lens}} \left(FK/kD \right) \right\} = \int_{-\infty}^{\infty} I(x) e^{iKx} dx \qquad (5.7)$$

where

x $\quad=\quad$ the coordinate in the image plane of the receiver,
k $\quad=\quad 2\pi/\lambda$,
K $\quad=\quad$ spatial frequency in radians per second,
D $\quad=\quad$ aperture diameter of the receiver,
M_L $\quad=\quad$ the MTF of the lens, and is given by

$$M_L\left(y \right) = \left(2/\pi \right) \left[\cos^{-1} y - y\sqrt{\left(1-y^2 \right)} \right], \text{ for } y \leq 1 = 0, \text{ for } y > 1 \ (5.8)$$

When the right side of (5.7) is determined experimentally, then the ATF of the atmosphere, $M_{\text{atmosphere}}$, is obtained as

$$M_{\text{atmosphere}} \left(FK/k \right) = M_{\text{lens}}^{-1} \left\{ \int_{-\infty}^{\infty} I(x) e^{iKx} dx \right\}_{\text{experimental}} \qquad (5.9)$$

The quantity $\rho_o = FK_o/k$ is determined from $M_{\text{atmosphere}}$ (FK / k) = 1/e.

However, since $M_L = 0$ for FK /kD > 1 , the method described above is applicable only for $\rho_o < D$, which is the case of interest here, because this condition is indicative of degradation of the beam by the turbulence in the atmosphere.

The measurements should be repeated at different wavelengths in order to facilitate checks and comparison with existing theory that shows that there is wavelength dependence on the beam spread, scintillation, angle of arrival and other turbulence parameters.

5.6 Methods of Obtaining a "Reference" Downlink Signal for Adaptive Optics Subsystem

Several methods exist which introduce a downlink signal whose purpose is to provide a reference signal for the adaptive optical subsystem located in the ground-based telescope antenna or in the telescope in an airborne platform, such a fixed or rotary wing aircraft, or even on the UAV platform. The first approach may be the use of a star as the reference. The second would be the downlink beacon from the spacecraft in the communication link. The third would be a dedicated downlink reference laser onboard the satellite that is part of the transceiver design. The fourth method would be a reference signal emanating from a laterally deployed secondary source to a mirror and from there reflected at the appropriate angle down to the earth station. The fifth option is an "artificial guide star" created by exciting the sodium atoms at about a 90-km altitude, by a ground-based laser. The sodium fluorescence emits downward radiation and is the reference that is used by the adaptive optical subsystem to make the necessary adjustments for use by the uplink.

In all of the above methods, the process involves measuring the distortion suffered by the downlink signal and then making adjustments of tiny mirrors in an array of mirrors making up the deformable mirror. These tiny mirrors move in different orientations to maximize the signal photons and minimize the tilt and phase distortion of the downlink wavefront of the reference signal. Having completed all the necessary adjustments of the deformable mirror, we then transmit the information-modulated laser signal back to the satellite, by way of the just-attained pattern of the deformable mirror. As the signal continues upward it goes through the atmospheric pathway through which the downlink just went and whose phase front was corrected. In this way, the uplink reaches the satellite receiver with a minimum of distortion.

The entire field of astronomy now uses AOS to substantially reduce the distortion of the stars' images. Before the development of AOS, the image of a typical star was hazy and poorly defined in terms of its twinkling and shimmering boundaries. But with AOS, the shimmering, twinkling, and haze are virtually eliminated and the boundaries become well detailed. However, the major problem with the laser receiver on the ground is the fact that the star is very far away and its light is only imaged within a small isoplanatic angle, when the satellite is also within view of the ground station.

However, for laser communications, using the space-based mirror to direct a reference downlink, as shown in Figure 5.5, has special advantages because the mirror does not weigh very much and is relatively easy to orient. Also the lateral reference source can be used to direct references to a number of optical ground stations. Alternatively, the reflected signal may be the desired

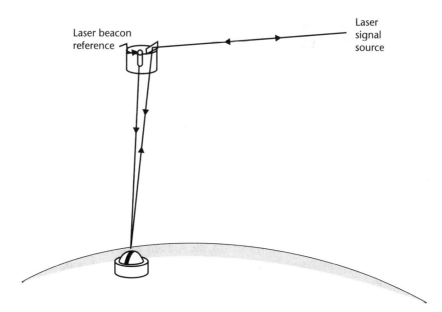

Figure 5.5 Use of the beacon laser to provide a reference to the AO subsystem on the ground, while the laser signal arrives laterally from a spatial source, which reaches the ground station via mirror reflector.

communications and the onboard laser source may be the dedicated reference downlink.

5.6.1   Synthetic Sodium Laser Beacon

The synthetic laser beacon concept is based on the excitation of sodium atoms in the mesospheric layer located in a 15-km wide layer at about 90-km altitude, by means of a ground based laser. Before the development of the ground-based laser reaching the 90-km region, there were experiments that excited the Rayleigh backscatter at altitudes of 6 to 20 kms. But because of the low-altitude excitation, the backscatter radiation generated poor focal anisoplanatism. By comparison, at the 90-km altitude excitation of the sodium gas, it turned out to be much more useful, with improved focal anisoplanatism.

The layer at 90 km was radiated with a dye laser having the following characteristics: wavelength of 0.589 µm; energy of each pulse, 40 m-j/GHz; pulse length, 4 µsec; pulse repetition rate, 20 pulses per second; and bandwidth, 3 GHz.

The test configuration [5] for the sodium reference laser system is shown in Figure 5.6.

As was observed, the data for the artificial beacon exhibits a higher level of photon noise and a relatively low-level signal from the sodium layer. In

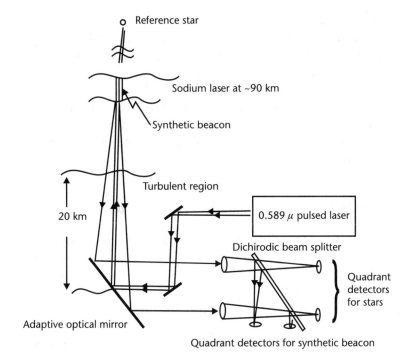

Figure 5.6 Exciting a sodium laser by a ground-based dye laser, which becomes the reference laser for the AO system of the ground station.

one experiment apart from the differences in noise levels, the beacon and the star (Beta Gemini) data was almost indistinguishable. However, the key here is that the artificial reference can be used anywhere, particularly when a star is unavailable.

5.7 Using a Reference Laser and an Oriented Mirror [5]

Deploying an oriented mirror on a relay satellite to reflect a laser signal from a primary high-data-rate signal source to the optical ground station (OGS), together with a reference laser on the same platform providing a downlink reference to the same OGS, will enable phase corrections for the uplink signal. Thus, adding a dedicated reference laser on the relay satellite platform, wherein a mirror on the same platform reflects the data signal from (for example) a LANDSAT type of satellite will also enable command messages to be sent to this LANDSAT from the OGS.

To stress the importance of the reference, it will be used by the AOS to help in applying compensatory distortions to cancel the optical distortions that are introduced by the atmosphere.

For an evaluation of a downlink signal, it may be desirable to look at values that others have developed for similar links. In [4], it was observed that even during the day the beam from the laser at synchronous altitude with $\lambda = 0.53$ μm and output power of 1.0 milliwatt with a diameter of 10 cm, generates sufficient signal to be collected with margin at the ground station having a 1.0-m aperture, with a common bandpass optical filter. Feasibility of a laser downlink with the AOS is then more than adequate to achieve closing of the link, with accommodation of broad bandwidths.

5.8 Uplink Signal Loss when AOS Is Not Used [6]

It has been shown that the effect of the atmospheric turbulence can be very deleterious in increasing the distortion of the uplink optical beam and consequently increasing the BER of the link. Of course, another source of the uplink and downlink losses that needs to be added to the loss evaluation is the weather (clouds, fog, rain, and snow). Its effects on the signal will be covered in Chapter 6.

Now, however, for purposes of illustrating the importance of having the AOS in the laser link, calculations are made of the uplink signal loss when the adaptive optics are not present in the link. It will be shown that the signal losses can be very large, making it not just desirable but mandatory to enhance signal performance by means of an adaptive optics subsystem in the telescope antenna.

Starting with a ground-based station with telescope that is located on top of a tower (Figure 5.7), at height $(Z_1 - Z_T)$, and is aimed at the satellite located at height Z_2, we have the height levels and associated angle as follows:

Z_1 = top of the tower, above sea level, on which the telescope antenna is perched

Z_2 = the vertical height of the satellite

Z_T = the height above sea level of the bottom of the tower

α = angle between the line of sight from the center of the aperture of the telescope of the satellite viewing the tower telescope to Z_2

Z' = running variable from Z_1 to Z_2.

Based on this we may write for ρ_o, the lateral coherence distance of a point source at the satellite observed at the ground transmitter, the following:

$$\rho_o = \left\{ 1.45k^2 \sec\alpha \int_{z1}^{z2} C_N^2\left(Z'\right)\left[\left(Z_2 - Z'\right)\right]^{5/3} dZ' \right\}^{-3/5} \qquad (5.10)$$

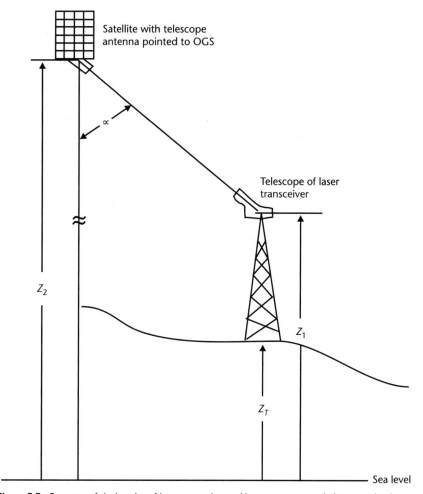

Figure 5.7 Geometry of the location of laser transceiver and its support tower relative to sea level.

where the key parameters are $C_N^2(Z')$, atmospheric index structure profile constant:

$$C_N^2(Z') \sim 10^{-6}\, d(z')/d_O \cdot \left[(2.88.1 + 1.34\lambda^{-2})/T(z')\right] C_T(Z'-Z_1) \quad (5.11)$$

$C_T(Z')$ = temperature Index structure constant,
$d(Z')$ = air density at altitude Z',
d_O = density of air at sea level.

The power loss due to the beam spread = $1/\theta_N^2$, where

$$\theta_N = \theta_T / \theta_o = \sqrt{\left[1 + \left(D/\rho_o\right)^2\right]} \qquad (5.12)$$

and

θ_N = normalized beamwidth relative to θ_o,
θ_o = diffraction-limited beamwidth in the absence of turbulence,
θ_T = beamwidth in turbulence.

The key in all these cacluations is the indication of distortion due to the turbulence, which is derived from (5.12).

The calculated uplink mean power loss due to turbulence is plotted in Figure 5.8. Plotted are the signal loss versus telescope tower height, above-ground site for average conditions on clear summer day and average conditions

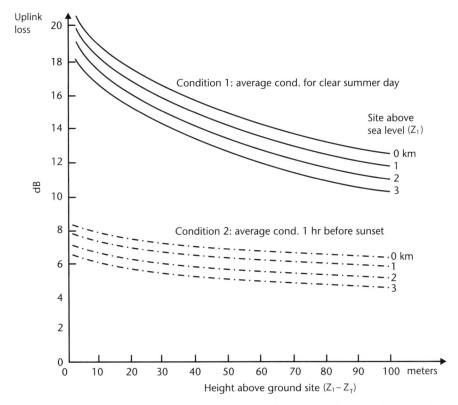

Figure 5.8 Uplink laser loss due to turbulence, when not using the adaptive optics system in the ground station.

1 hour before sunset. These were calculated at sea level, then 1 km, 2 km, 3 km above sea level. As seen, on a clear summer day at 1 km above sea level, a 20-dB loss was sustained when the tower height was very low, as great as ~3m. A 10-dB loss was sustained when the tower height was 100m, and its site was 3 km above sea level. At 1 hour before sunset, the result was less than half of the signal loss of that of a clear summer day. In general the loss varies with the conditions of weather, temperature, and time of day.

5.9 Summary and Concluding Remarks

This chapter shows by means of simple algebraic equations how the signal distortion on the uplink, expressed by signal coupling coefficient, Γ_T', can be estimated by the inverse square ratio of D/ρ_o where D diameter of the aperture and ρ_o is the lateral coherent distance (Fried's Length). That is,

$$\Gamma_T' \sim 1/\left\{1+(D/\rho_o)^2\right\}$$

Thus, Γ_T' becomes a factor in the SPB between the laser transmitter and receiver (given in Chapter 2). As indicated, when $D > \rho_o$ there is a loss of photons per bit.

Further shown by measurements, C_N^2, the atmospheric structure constant may be obtained by measurement and then may be used to calculate ρ_o, which enables the evaluation of the signal coupling coefficient. It is stressed in Chapter 5 that an adaptive optical system is desired for every laser ground station and every airborne station. Such a design will help to improve the performance of the communications link, as often the atmospheric conditions vary sufficiently to cause ρ_o to be smaller than D, even if at certain times of the day ρ_o is larger than D. That is, the AOS is needed to accommodate the occurrences of any condition of turbulence.

References

[1]   Tyson, R. K., *Principles of Adaptive Optics*, Philadelphia: Academic Press, 1991.

[2]   Sasiela, R. J., *Electromagnetic Wave Propagation in Turbulence*, New York: Springer Verlag, 1994.

[3]   Culman, D. K., *Journal of Solar Physics*, Vol. 7, 1969, pp. 122–129.

[4]   Humphreys, R. A., C. A. Pimmerman, L. C. Bradley, and J. Herrmann, "Atmospheric Turbulence Measurements Using a Synthetic Beacon in Mesospheric Sodium Layer," *Optics Letters*, Vol. 16, No. 18, Sept. 15, 1991, pp. 1367–1369.

[5]   Armstrong, J., C. Yeh, and K. Wilson, *Correcting for Atmospheric Effects in Optical Communications,* Cal Tech–JPL-NASA Report NPO-20506, Sept. 2004.

[6]   Yura, H., "Uplink Mean Power Loss Due to Turbulence," lecture given at Aerospace Corporation, summer 2002.

6

Terrestrial Laser Communication Links and Weather Issues

6.1 Introduction

In the process of transmitting a laser signal beam from one point to another point over a terrestrial link, which might be typically 30 km long and 40m above the ground, we encounter a number of signal-loss components. These include absorption and scattering by airborne molecules and aerosols and wavefront distortion due to atmospheric turbulence resulting from the variation of the index of refraction along the beam's path.

Although a 30-km length is chosen as a typical example, there are larger link distances between the laser transmitter and the laser receiver. For example, up to 148 km was achieved in the Canary Island experiment, but with tall towers or mountain ridges supporting the laser antenna telescopes.

Expressions for the correction of the distortion will be presented in this chapter, as well as a list of the losses due to absorption and scattering in clear weather and also with weather conditions such as rain, fog, clouds, and snow.

Because of the very large losses that can be caused by weather conditions, a couple of ameliorative approaches are introduced, which come under the weather avoidance system (WAS) architecture. These include diversity reception, which interlinks OGSs via underground fiber cable nets, and even an OGS perched on the top of an airship that is hoisted up to several kilometers in height, as well as by means of a mechanical wire twisted with a fiber cable that will connect the top-mounted telescope platform with the associated ground-based transceiver components. The height of the airship may often be sufficient to avoid most weather issues. The height of the balloon hoisting the telescope would be of the order of 4 km.

However, in the case of a WAS system, its major advantage relies on the ubiquity of underground and/or overhead fiber cables, which can be easily integrated with any number of OGS stations in any geographical area. Thus, a downlink signal coming to one OGS in dry weather will be able to have that laser signal transmitted to locations that exist in inclement weather by means of the underground fiber cable. In this manner, the basic integration of all OGS and communication substations in any weather should be feasible. The connecting fiber nets may be landline or submarine cables. Thus, the entire world is, in principle, connectable by means of integrated OGSs and fiber cables.

As an introduction to the need for the WAS, we will proceed in the next section with expressions for the atmospheric losses, starting with signal distortion due to atmospheric turbulence, followed by atmospheric attenuation due to molecular and aerosol scattering and absorption, and ending with laser signal loss due to haze, fog, rain, and snow.

6.2 Calculations of Atmospheric Turbulence Parameters

A discussion of how to calculate the basic equations of turbulence is summarized below. Several rules of thumb are presented that are useful, when the particular levels of $C_N^2(h)$, the atmospheric structure constant, and $\rho_{o,}$ the corresponding coherence diameter, are given, in determining the size of aperture necessary to attain a reduction in the distortion. Depending on the ratio of the aperture diameter to the coherence distance, we determine whether it is necessary to integrate an AOS into the design of the optical system of the laser transceiver.

Starting with the Hufnagle-Valley-Bufton [1] atmospheric turbulence model,

$$C_N^2(h) = 0.00594 (W/27)^2 (10^{-5} \cdot h)^{10} \exp(-h/1000) + 2.7 \cdot 10^{-16} \quad (6.1)$$
$$\times \exp(-h/1500) + A \exp(-h/100)$$

For $A = 1.7 \cdot 10^{-14}$, $W = 21$ mi/sec, and h = 40m, (6.1) becomes

$$C_N^2(40) = 1.15 \cdot 10^{-12} \, \text{meters}^{-2/3} \quad (6.2)$$

And for the distance of 30,000m, the 0th turbulence moment, μ_o, for 0° zenith angle may be evaluated directly from

$$\mu_o(L) = \int_o^L dz C_N^2(40) = \{30,000\}\{C_N^2(40)\} \tag{6.3}$$

In addition, ρ_o, the coherence diameter, can be obtained as shown:

$$\rho_o = \{0.423 \, k^2 \mu_o\}^{-3/5} \tag{6.4}$$

The need for the AOS is again determined by the expression for Γ_T, the turbulence signal coupling efficiency,

$$\Gamma_T \sim 1/\{1 + (D/\rho_o)^2\} \tag{6.5}$$

Thus, the ratio of $D/\rho_o << 1$, indicates little distortion and therefore does not require an adaptive optics subsystem. Alternatively, when the reverse is true, and $D > \rho_o$, then AOS is quite useful. Moreover, it is necessary to multiply the SPB, which is derived in Chapter 2, by the factor Γ_T, as expressed in (6.5). The SPB will thus be an effective measure of the number of photoelectrons per bit, when facing turbulence in the atmosphere.

6.3 Absorption and Scattering in the Atmosphere

The molecular scatter and absorption constants and the aerosol scatter and absorption constant may be added to produce the basic attenuation constant, γ, for the wavelength of the beam that is propagated through the atmosphere. Its attenuation is expressed through Beer's Law as the following two equations:

$$T_R = I_Z / I_o = \exp(-\gamma z) \tag{6.6}$$

where

T_R = transmittance
I_o = beam intensity at the start of its journey from the telescope
I_z = beam intensity at distance z from the telescope

$$\gamma = \alpha_M + \beta_M + \alpha_A + \beta_A \tag{6.7}$$

where

α_M = molecular absorption constant
β_M = molecular scattering constant
α_A = aerosol absorption constant
β_A = aerosol absorption constant

For the molecular absorption constant, when the impinging wavelength is greater than the size of the molecules and when those molecules are primarily composed of H_2O and CO_2, the most prevalent absorbers, the total molecular absorption is gotten by summing over each of the molecular types and their allowed transitions.

The molecular scattering constant is derived by using the second order differential equation, which describes the induced dipole under the application of a harmonic field. As discussed by Hugo Weichel in [2], it is equal to

$$\sigma = f'e^4\lambda_o^4 / 6\pi\varepsilon_o^2 m^2 c^4 \lambda^4 \qquad (6.8)$$

where

- f' = oscillator strength; the effective number of electrons per molecule that oscillates at the natural frequency ω_o. The maximum value of the oscillator strength is equal to the total number of electrons in the molecule. (The scattering cross-section as shown above is known as the Rayleigh Scattering.)
- e = charge of electron
- λ = wavelength of the laser beam
- λ_o = wavelength gotten from ω_o. That is, $2\pi f_o = 2\pi\ c/\lambda_o$ and $\omega_o = \sqrt{(k/m)}$, the natural frequency of the molecule

In terms of Mie Scattering, where the size of the particles is on the order of the impinging wavelengths, we have an encounter between the laser communication beam and the small water droplets and aerosols, as expressed in Chu and Hogg [3], and take into consideration the size, shape, density, composition, dielectric constant, and absorptivity of the particles. As it turns out, a particle with the same product, rk, whereby its radius, r, and its propagation constant, k, has the same scattering characteristics. Now by applying laser light intensity, I, of cross-sectional area A and wavelength, λ, into a small, tubular structure of length dz and volume $\pi a^2 dz$, the fractional decrease in intensity as the beam goes through the volume element is

$$-dI/I = K\pi a^2 NAdz/A \qquad (6.9)$$

where

- $-dI/I$ = fraction of laser intensity that is reduced when passing through elementary volume of size Adz
- $NAdz\ \pi a^2$ = the total cross-section of the particles

N   = total number of particles in the volume element interact-
               ing with the laser beam

K   = attenuation factor, which is due to the scattering and
               absorption of the particles whose size are similar in mag-
               nitude to the impinging wavelength.

Thus (6.3) may be rewritten as

$$-dI/I = N\sigma(a,\lambda)dz \qquad (6.10)$$

where

$N\,\sigma(a,\lambda)\,dz$ = Mie attenuation coefficient
$\sigma = K\pi a^2$ = Mie attenuation cross-section

It is inferred that the laser signal is reduced in intensity by the Mie process, due to scattering and absorption.

Measurements of aerosol scattering coefficients and the associated relative humidity have been made in many regions of the country and across the world, as a function of laser wavelength. An important empirical relation further developed by Hugo Weichel [2] for the scattering coefficient in the atmosphere, which considers both the Rayleigh (note the wavelength to the negative fourth power) and Mie Scattering, is

$$\beta(\lambda) = C_1 \lambda^{-\delta} + C_2 \lambda^{-4} \qquad (6.11)$$

where

λ = laser wavelength
C_1, C_2, and δ = constants determined by aerosol density (concentration)
               and distribution of the physical size of the particles

Since all the wavelengths are greater than 0.3μ, the second term of (6.5) is smaller than the first and may therefore be neglected. But the value of $\delta \approx 1.3 \pm 0.3$ has been estimated from the measurements. However, relating the C_1 and δ to meteorological parameters is often done by the U.S. Weather Bureau, and these are noted periodically to various research institutions.

By definition, the contrast of a laser signal source at 0.53μ viewed at a distance, z, may be expressed as

$$C_z = (R_{sz} - R_{bz})/R_{bz} \qquad (6.12)$$

where

C_z = contrast of the laser signal source as viewed from a distance of z kilometers

C_z = contrast of the laser signal source as viewed from a distance of z kilometers

R_{sz} = laser source that is observed from z kilometers away

R_{bz} = radiation background that is observed from z kilometers away

The ratio of the contrasts, at distance z, relative to distance 0, is defined as the visual range; thus,

$$C_z / C_o = (R_{sz} - R_{bz}) / R_{bz} \div (R_{so} - R_{bo}) / R_{bo} = 0.02 \qquad (6.13)$$

If it is further assumed that the signal is much more intense than the background and that the background is constant, (6.13) may then be written as

$$R_{sv} / R_{so} = \exp(-\beta V) = 0.02 \qquad (6.14)$$

where

V = visual range

β = scattering coefficient

R_{sv} = signal radiation observed at visual range, V

R_{so} = signal radiance when observed very close to the source

Equation 6.8 may also be written as a function of the natural logarithm, ln:

$$\ln[R_{sv} / R_{so}] = -\beta V = -3.91 \qquad (6.15)$$

And from (6.5) we have

$$\beta = 3.91 / V_{km} = C_1 \lambda^{-\delta} \qquad (6.16)$$

which, for wavelength of the laser signal as 0.53μ, is

$$C_1 = 3.91 / V_{km} (0.53)^{\delta} \qquad (6.17)$$

Furthermore, the transmittance at the center of the ith window is

$$\tau_i = \exp\left[-3.91 / V_{km} (\lambda_i / 0.53)^{-\delta} Z\right] \qquad (6.18)$$

where

Z   =   distance along the visual range in km from the source to the observing sensor

V_{km} =   visual range in km

λ   =   wavelength in microns within the ith window

It should be noted that for outstanding visibility, the power level, δ, is 1.6 and for average visibility the power level is 1.3. Therefore, we can compute the transmittance the instant we know the visual range and also by knowing the relative humidity. In fact, the coefficient for extinction that is equivalent to the coefficient for absorption and scattering is plotted in Figure 6.1 for a 23-km visibility. As we see, at 1μ the attenuation coefficient, based on the total aerosol extinction, is ~ 10^{-1}/km; at a 0.5μ wavelength, the attenuation coeficient is ~ $2 \cdot 10^{-1}$/km. At a 10.6μ wavelength, the attenuation coefficient is ~ $2 \cdot 10^{-2}$/km.

To continue, when calculating the propagation though weather particulates such as haze, fog, and rain, the scattering coefficient, β, may be expressed as

$$\beta = 1.25 \cdot 10^{-6} \left(\Delta x / \Delta t \right) a^3 \tag{6.19}$$

where raindrops are larger than the wavelength; therefore, there is no wavelength-dependent scattering:

$\Delta x / \Delta t$ =   rate of rainfall in centimeters of depth/sec

a   =   raindrop size in centimeters

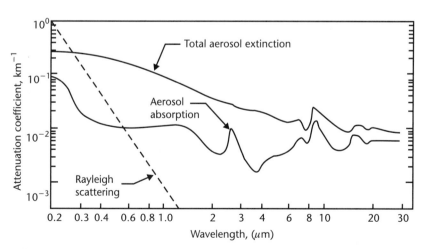

Figure 6.1 Aerosol absorption and extinction coefficient for 23-km visibility and a continental aerosol model [2].

Table 6.1

Transmittance of a 1.8-km Path Through Rainfall of Various Intensities [3]

Condition	Rainfall	Transmittance of 1.8-km Path
	$\Delta x / \Delta t$ in cm/hr	
Light rain	0.25	0.88
Medium rain	1.25	0.74
Heavy rain	2.5	0.65
Cloudburst	10.0	0.38

As an example, the transmittance in a cloudburst with a rainfall rate of 10 cm/hr ($2.77 \cdot 10^{-3}$ cm/sec), and with raindrop radii of 0.025 cm to 0.325 cm, is presented in Table 6.1. As seen, for a distance of 1.8 km, the transmittance is in the range of 0.88–0.38.

Finally, the scattering coefficient based on a cloudburst for rain at 10 cm per hour under the explicit conditions of drop radii and concentration of drops per cm^2 is given in Table 6.2. To estimate the signal attenuation, one uses the scattering coefficient in the Beer's Law expression.

Table 6.2

Calculations of the Scattering Coefficient for a Cloudburst Condition of Rain at a Rate of 10 cm/hr [4]

Drop Radius (cm)	Number of Drops per cm2 of Horizontal Area in 100 Seconds	Rainfall Rate cm/sec	Scattering Coefficient (cm−1)
0.025	43	$2.78 \cdot 10^{-5}$	$2.24 \cdot 10^{-6}$
0.05	21.4	$1.11 \cdot 10^{-4}$	$1.11 \cdot 10^{-6}$
0.075	14.3	$2.5 \cdot 10^{-4}$	$7.43 \cdot 10^{-7}$
0.10	9.3	$3.89 \cdot 10^{-4}$	$4.87 \cdot 10^{-7}$
0.125	5.8	$4.72 \cdot 10^{-4}$	$3.02 \cdot 10^{-7}$
0.150	3.6	$5.00 \cdot 10^{-4}$	$1.865 \cdot 10^{-7}$
0.175	1.8	$4.07 \cdot 10^{-4}$	$9.59 \cdot 10^{-8}$
0.200	0.75	$2.50 \cdot 10^{-4}$	$3.91 \cdot 10^{-8}$
0.225	0.35	$1.67 \cdot 10^{-4}$	$1.83 \cdot 10^{-8}$
0.250	0.13	$8.36 \cdot 10^{-5}$	$6.76 \cdot 10^{-9}$
0.275	0.064	$5.56 \cdot 10^{-5}$	$3.34 \cdot 10^{-9}$
0.300	0.024	$2.78 \cdot 10^{-5}$	$1.29 \cdot 10^{-9}$
0.325	0.019	$2.78 \cdot 10^{-5}$	$1.02 \cdot 10^{-9}$
	$\Sigma = 100.54$	$\Sigma = 5.234 \cdot 10^{-6}$	

6.4   Attenuation Due to a Variety of Weather Components

The signal attenuation due to absorption and scattering in clear weather, haze, light and heavy fog, light and heavy rain, and light and heavy snow is shown in Table 6.3. This table is of particular value since the loss extrapolation is extended from distances of 1–100 kilometers [5, 6].

As a further rule of thumb, the effect of a laser beam at 0.53μ penetrating a cumulus cloud containing liquid water concentration of 1.77 gm/m^3 and thickness of 230m will be to produce an attenuation of 17 dB. The same cloud characteristics and a thickness of 677m will attenuate the laser signal at 50 dB, and a cloud thickness of 2,100m will attenuate the signal at 156 dB. The data source for this information is Subramanian [5].

The experimental measurements shown in Table 6.3 emphasize the large losses suffered by the laser signal when going through haze, fog, rain,

Table 6.3

Estimated Attenuation Due to Absorption and Scattering Caused by Weather Parameters for 1-, 10-, and 100-km Terrestrial Links [5, 6]

Weather	Wavelength, λ	Attenuation in dB at L Distance		
		1 km	10 km	100 km
Conditions	microns			
Clear weather (at sea level)	0.53, 1.06	0.06	0.6	6
	10.6	0.54	5.4	54
CO_2 absorption	0.53, 1.06	–	–	–
	10.06	0.25	2.5	25
Haze	0.53, 1.06	1.4	14	140
Size; 0.1 mg/m³	10.6	0.66	6.6	66
Light Fog (0.5-10μ size;	0.53, 1.06	0.1-5	1-50	10-500
0.5 mg/m³; visibility ~2 km)	10.6	0.9	9	90
Fog (0.5-10 μ size; 1 mg/m³	0.53, 1.06	0.2-10	2-100	20-1000
visibility ~0.5 km)	10.6	1.9	19	190
Rain: 5mm/hr	0.53, 1.06	1.6	16	160
25mm/hr	0.53, 1.06	4.2	42	420
75mm/hr	0.53, 1.06	7.0	70	700
Light rain (1000μ size; 50 mg/m³)	10.6	1.6	16	160
Snow: Light	0.53, 1.06	1.9	19	190
Heavy	0.53, 1.06	6.9	69	690

and snow. These loses are calculated from a number of software systems prepared by the Air Force Geological Laboratory (AFGL) at Hanscom Air Force Base. This data can help to estimate the range between laser repeater stations as well as give us a measure of the BER. Importantly, it is the major challenge that laser communications faces because the laser links must at times operate in inclement weather. Another piece of AFGL software relevant to the estimation of laser signal loss is outlined in the next section.

6.4.1   MODTRAN System for Estimating Laser Signal Penetration of the Atmosphere

An additional software system has been developed by the Air Force Research Laboratories (AFRL), Space Vehicle Directorate, together with Spectral Science, Inc., which provided the atmospheric Radiative Transfer Code System and algorithms. The title of the system is MODTRAN. It is an approximate abbreviation of Moderate Spectral Resolution Atmospheric and Transmittance Algorithm and Modeling.

The code calculates the transmission and radiance for frequencies from 0 to 22,680 cm^{-1} with a resolution of 2 cm^{-1} and between 22,680 and 50,000 cm^{-1} with resolution of 20 cm^{-1}. This code is based on the work of the earlier system known as LOWTRAN, and it covers spherical refractive geometry, solar and lunar background sources, with Rayleigh, Mie, single and multiple scattering and default profiles such as due to gases, aerosols, clouds, and rain.

Other features of the MODTRAN software are penetration of laser signals, in the visible and infrared, using Beer's Law modeling with six climatological descriptions. These are tropical, midlatitude, summer/winter, and subarctic summer/winter. Also included are U.S. standards for six atmospheric gases, H_2O, CO_2, O_3, N_2O, CO, and CH_4, and single profiles for HNO_3, NO, NO_2, SO_2, O_2, N_2, NH_3, and the heavy molecules.

The aerosol profiles include tropospheric, rural, urban, desert, sea, and fog, plus the stratospheric, which includes volcanic (background, aged, high, fresh, and extreme) and the clouds and rain covering cumulus, altostratus, stratus, stratocumulus, nimbostratus, and cirrus (standard, subvisual, and NOAA). Also included are geometric lines of sight H1 (observer location) to H2 (end of path) with H1 or H2 as the surface, space, or any place within. The radiation sources are the local thermodynamic equilibrium (LTE) thermal and surface radiation and solar or lunar irradiance. From these properties it may be seen that MODTRAN may be used to predict the laser signal radiation transmittance through the atmosphere and in most weather conditions along a terrestrial path.

6.5 The Weather Avoidance System

Because weather attenuates laser communications in the atmosphere, the initial approach that is used is to get around the weather problem is by trying to locate optical ground stations in very dry regions of the country. We started by locating the optical ground stations in the Southwest region of the United States. Similar regions exist in other parts of the world. The potential locations that are appropriate for building our ground stations, from a weather isolation aspect, are discussed in the next section.

6.5.1 Examples of Dry Weather Locations in the Southwest Region of the United States

Shown in Figure 6.2 are the estimated dry days at three selected sites: Black Top Mountain, New Mexico, which averages 94.4% of clear days—defined as days with less than 0.1 inch of precipitation—per year; Kingston Peak, California, averaging 95.5% clear days; and Panamint Range, California, averaging 97.3% clear days.

Additional locations in the Southwest portion of the United States are given in Table 6.4. Twenty-three locations are mentioned. Apart from the level of dryness, there are aspects of physical accessibility to the location from the main transportation roads and passages. We list these and other locations in the United States in which we are able to connect the stations to one to another via a ground-based laser fiber network, thereby enabling us to achieve the equivalent of a cloud-free line of sight.

Table 6.4

List of Potential Locations of Laser Ground Stations Based on Dry Weather and Accessibility of Transportation

1. Apache Mountain, NM	13. Luera Peak, NM
2. Apache Peak, AZ	14. Millers Peak, AZ
3. Atascosa Peak, AZ	15. Mt. Wrightson, AZ
4. Baldy Peak, AZ	16. Nogal Peak, NM
5. Big Hatchet Peak, NM	17. Oscura Peak, NM
6. Black Top Mountain, NM	18. Panamint Range, CA
7. Capitan Mountains, NM	19. Rose Peak, AZ
8. Capitol Peak, NM	20. Sacramento Mountains, NM
9. Chiracahua Peak, AZ	21. Salinas Peak, NM
10. Emory Peak, TX	22. San Andreas Peak, NM
11. Guadalupe Mt. Range, NM	23. Sierra Blanca, NM
12. Kingston Peak, CA	

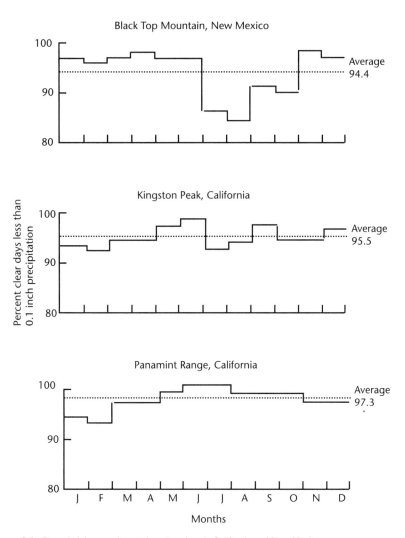

Figure 6.2 Recorded dry weather at three locations in California and New Mexico.

Various combinations of ground stations may be considered to enhance our weather avoidance strategy. Shown in Figure 6.3 is an example in which two ground stations are interrelated to accommodate their weather adversity, that is, shifting to the dry location when the other geographical spots become weather-covered and vice versa. Although we show only two stations, one at the Panamint Range in California and the other in Black Top Mountain in New Mexico, more than these two may be integrated into the set of stations that are selectable for dry locations. In order to determine which are the dry ones when

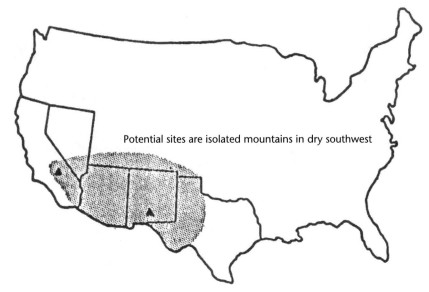

Potential sites are isolated mountains in dry southwest

Use of several sites with proper weather cross correlation can achieve >99% downlink availability

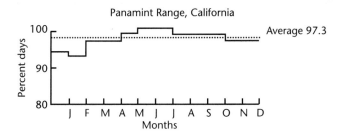

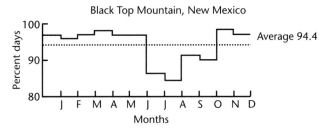

Figure 6.3 Interspaced locations of two optical ground stations at which dry weather occurs at different times of the year.

a particular Earth link is required, we employ data derived from weather satellites through NOAA and other weather sources.

Another example of a set of ground station locations that may be combined through optical fiber cables and be selected for clear weather advantage, throughout the United States, but different from locations in the Southwest, is shown in Figure 6.4. The six ground stations are LND (Lander, WY), YUM (Yuma, AZ), AMA (Amarillo, TX), EVV (Evansville, IN), DCA (District of Columbia), and TPA (Tampa, FL). Because this set has broad geographical coverage, we are able to consider the probability of a cloud-free line of sight (PCFLOS), and if we consider one out of two or one out of three stations, we get a very large PCFLOS.

Table 6.5 notes that the PCFLOS for three out of the six stations previously (YUM, LND, and AMA), in the months of January, April, July, and

Table 6.5

Probability of Cloud-free Line of Sight when Considering Two and Three out of Six Network Stations

Selected Combinations of Network Stations	Probability of Clear Line of Sight (Percent)				
	January	April	July	October	Annual
1. YUM	83	94	92	93	91
2. LND	66	66	76	67	69
3. AMA	71	75	81	76	76
4. EVV	46	65	82	73	65
5. TPA	63	74	61	67	68
6. DCA	46	57	64	61	58
1 & 2	94	98	98	98	97
1 & 3	95	99	99	98	98
2 & 3	90	91	95	92	91
2 & 4	82	88	96	91	89
3 & 4	84	91	97	93	91
3 & 5	89	93	93	92	92
4 & 5	80	91	93	91	91
4 & 6	71	85	94	89	86
5 & 6	80	89	86	88	87
*1, 2 & 3	98	99+	99+	99	99
*2, 3 & 4	95	97	99	98	97
*3, 4 & 5	93	98	99	98	97
*4, 5 & 6	93	96	97	96	95

Figure 6.4 Proposed locations of optical ground stations spaced throughout the continental United States.

October, is 99%. Other high-probability values are also attainable when considering, for example, one out of seven stations' locations across the width of the United States from Ft. Yukon, AK, to Portland, ME. The PCFLOS is calculated and presented in Table 6.6.

As shown, the probability of one out of seven stations turns out to be high: 0.9995 based on the value of each station having PCFLOS of 0.499 to 0.784. This means that a satellite may select one out of seven ground stations, based on weather data that it receives from NOAA, Defense Meterological Satellite Program (DMSP), Geostationary Operational

Table 6.6

Probability of at Least One Site Having a Cloud-free Line of Sight out of n Independent Sites
(The Product Symbol Extends from i = 1 to i = n.)

Site	Location	P^i_{cflos}	$P_n(1) = 1 - \Pi^n_{i=1}(1 - P^i_{cflos})$
1	China Lake, CA	0.784	$P_1(1) = 0.784$
2	White Sands, NM	0.697	$P_2(1) = 0.935$
3	Kahului, HI	0.658	$P_3(1) = 0.978$
4	Denver, CO	0.623	$P_4(1) = 0.992$
5	Winnemucca, NV	0.621	$P_5(1) = 0.997$
6	Portland, ME	0.499	$P_6(1) = 0.998$
7	Ft. Yukon, AK	0.665	$P_7(1) = 0.9995$

Environmental Satellite (GEOS) program, and the National Polar-Orbiting Operational Environmental Satellite System (NPOESS) and other data-integrating sources. Thus, the downlink from a satellite to the Earth station will be aimed at a location that has a very minimum of cloud cover. And from that station, the data will be transmitted to other locations in the country via fiber cable. In this way, any location within the United States and, to a smaller degree, elsewhere in the world that has fiber cable connectivity to a U.S.-based station, will be connected to a laser downlink from a satellite. Fiber cable network have been laid by the Baby Bell companies in the United States and other phone companies, including Global Crossing, have also laid submarine cables under the major oceans of the Earth. This makes the broadband capability of one ground station capable of reaching other stations at different locations of the Earth in which the weather may be inclement. Interestingly, most of the fibers of the cable network, at this writing, are "unlit," so that when the time comes for their utility, the fibers are very likely to be available.

There are additional features associated with the selection of the location of the ground stations. As we have said, they have to do with the fact that weather data used in the selecting of an OGS may also be used in the selection of terrestrial stations. This process involves locating OGS at dry locations, building towers of sufficient height to be above the large thermal turbulence that is typically close to the ground, and transmitting terrestrially to another tower through the atmosphere. Communications links between the terrestrial towers may be integrated via underground or overhead cable gateways into a national fiber network.

6.5.2   Pictorial Representation of the Weather Avoidance System

Figure 6.5 presents a sketch showing the selection of a ground station location that is useful as an Earth station for satellite laser communications and also as a station for terrestrial communications.

6.6   Testing of Laser Communications Along Terrestrial Links

There have been two major measurement programs that have supported terrestrial links projects. The first is the 148-km link in the Canary Islands and the second is a 28-km link in northern California.

The 148-km terrestrial link was the ground leg of the Semiconductor Laser Intersatellite Link Experiment (SILEX), carried out in the Canary Islands between two tall mountains, and a propagation path along the sea between the towers located at La Palma and the Tenerife islands. The laser

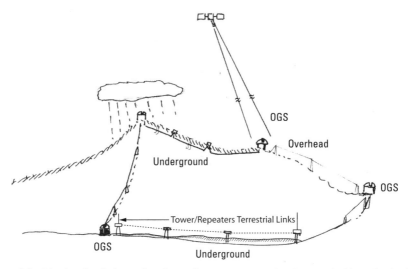

Figure 6.5 Selection of optical ground stations and tower repeater stations integrated with overhead and underground fiber cable.

wavelengths considered in the measurement program were 0.790μ, 0.870μ, 1.064μ, 1.3μ, and 10.2μ. The measurement effort covered the effects of absorption and scattering andintensity scintillations as well as turbulence measurements and the angle of arrival fluctuations.

A combination of the signal strength and the noise components [7] is shown in Figure 6.6, for the ground phase (148-km length) of the SILEX program. As seen, the shot noises due to the signal as well as clear sky, are small, even with the Sun in the FOV of the receiver, and the S/N is roughly 25 dB for the case of low atmospheric attenuation of 4.5 dB. It should be stressed, however, that there were no wideband signal measurements and no associated BER measurements for any selected modulation scheme. The emphasis in this effort was the measurements of a narrowband signal about the laser wavelength. Moreover, there was not any adaptive optical subsystem designed into the receiver optics as was the case in the measurement program developed by the LLNL [8].

In the LLNL measurement program, an adaptive optical system was used together with a beacon, in order to first measure the wavefront distortion and then make the necessary adjustments by the deformable mirrors to correct for the turbulence in the atmosphere. This approach reduced the BER of the wideband signal propagated along the terrestrial link. It should be noted that the deformable mirror was actually composed of a large number, initially of 11 × 11 mirror elements (with plans for 128 × 128 mirror elements) of microelectromechanical systems (MEMS). The MEMS system was

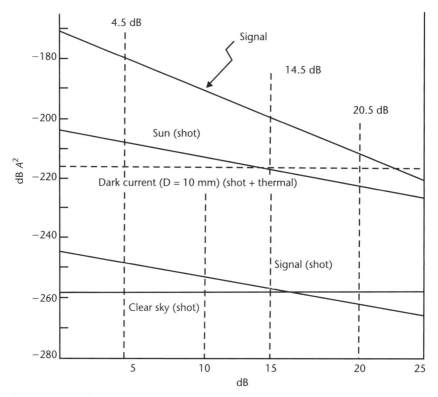

Figure 6.6   Signal power budget and shot noise for the 148-km terrestrial link [7].

the first to be used in other terrestrial link experiments. While ranges of 28 km were part of the set of test measurements, the emphasis was placed on the "last mile," or in fact the "last kilometer," type of wideband laser communication link design, with up to a 20-Gb/sec data rate.

The 28-km link was from the LLNL Laboratories to Mt. Diablo, while the shorter lengths, of 1.3 km, were part of the last mile configuration. These are especially useful in commercial telephony in providing wideband signals to businesses and homes, from fiber cables through the atmosphere and then to appropriate receivers in the subscriber establishments, at a 1.55μ wavelength.

6.7 Concepts for Penetrating Low Clouds Surrounding a Ground Station or Airstrip

Several concepts have been under development to provide a "hole" in a fog field or through a low dark cloud, to enable the penetration of a low-power laser communication signal through it. The process begins by vaporizing a hole through the cloud with a high-power laser. The magnitude and nature of the optical wavefront aberrations present in a laser-cleared hole in water-laden fog and cloud may be estimated. Under some atmospheric conditions and based on the atmospheric time constant, the vaporized cloud may recondense into highly scattering aerosols and thereby severely inhibit propagation through the hole. However, the period of time between the hole creation and recondensation generally enables the propagation of a "burst" of a high-data-rate signal, at a low attenuation. While this approach may be used, caution must be exercised to ensure that all persons in the area where the high-energy laser is transmitting wear special protective goggles. This is of course necessary to avoid retinal damage.

6.8   Summary and Concluding Remarks

The evaluation of signal loss of a terrestrial laser beam over distances of 28 km and 148 km was considered in this chapter, with extrapolation to longer distances using taller towers upon which telescope antennas were deployed. The losses considered include absorption and scattering by airborne molecules and aerosols and wavefront distortion induced by atmospheric turbulence.

The evaluation of atmospheric and weather loss is based on data available through particular measurements and also by the use of LOWTRAN and the more recent MODTRAN. Both software systems were developed under the management of the AFGL at Hanscom AFB. The MODTRAN program provides the transmission and radiance at the wavelengths of interest. MODTRAN may also be used to predict the laser signal radiation through the atmosphere, and in most weather conditions, along a terrestrial path. In the future, integration of near-real-time weather data from current and future weather satellites, such as DMSP, GEOS, NPOESS, and other satellites, with data-relay platforms may be undertaken. The weather data may be transmitted from a ground-based processing station, or directly from the weather observing satellites, enabling the laser communication platform to transmit the laser signal directly down to a selected weather-free ground station.

An ameliorative approach to bypass inclement weather conditions over a particular OGS was introduced in this chapter, called the WAS. It integrates the cable fiber network and the OGS. That is, the various ground stations are interconnected by the existing ground-based, overhead fiber cable, and even submarine cable. In this manner, if the intended OGS is affected by inclement weather, the ground station that has clear weather will receive the downlink communications, but the data will also be received by the weather-affected station, through the available fiber network. It is recognized that while the major lengths of the fiber cable that are used will be those portions of the net that are currently "unlit," there is likely to be a need to lay fresh cable, though of relatively short distance, to connect those ground stations that are not currently serviced by fiber cable.

Weather patterns have indicated that using only 1 ground station out of 10, for example, in a large land mass such as the United States, will present a large figure of PCFLOS, in excess of 0.999. Thus a transmitted downlink to the zone internal (ZI) may be assumed to be capable of closing the link with a high probability, particularly when the land-based fiber network is combined with the OGS. In other countries or a combination of countries, the same pattern of a high probability CFLOS will hold. However, the utility of a ground-based and overhead cable network and also the inclusion of a submarine cable may be required.

References

[1]   Hufnagle, R. E., and N. R. Stanley, "Modulation Transfer Function Associated with Image Transmission Through Turbulent Media," *Journal of the Optical Society of America,* Vol. 54, Jan. 1964, pp. 52–61.

[2]   Weichel, H., *Laser Beam Propagation in the Atmosphere,* Volume TT-3, Bellingham, WA: SPIE Optical Engineering Press, 1990.

[3]   Chu, T. S., and D. C. Hogg, "Effects of Precipitation on Propagation at 0.53, 3.5, and 10.6 micron," *Bell System Technical Journal,* Vol. 47, May–June 1968, pp 723–759.

[4]   Hudson, R. D., *Infrared Systems Engineering,* New York: John Wiley and Sons, 1969.

[5]   Subramanian, M., "Atmospheric Limitations for Laser Communications," *EASCON Record,* 1968.

[6]   RCA Labs, *Research Program on the Utilization of Coherent Light,* AD 276526, April 20, 1962.

[7]   Ruiz, D., R. Czichy, J. Bara, A. Cameron, A. Belmonte, P. Menendez-Valdes, F. Blanco, and C. Pedreira, "Inter-Mountain Laser Communication Tests," *Free-Space Laser Communication Technologies II,* SPIE, Vol. 1218, 1990, pp 419–430.

[8]   Wilks, S. C., J. R. Morris, J. M. Brase, S. S. Olivier, J.R. Henderson, C. Thompson, M. Kartz, and A. J. Ruggiero, "Modelling of Adaptive Optics-Based Free Space Communications Systems," *Proc. of SPIE,* Vol. 4281, 2002.

7

The Fifth-Generation Internet System

7.1 Introduction

This chapter discusses the architecture of interconnecting satellites with airborne and ground-based platforms via laser links and supported by a terabits-wide laser backbone, located at a synchronous altitude. We call this wideband communication system the Fifth-Generation Internet (5-GENIN). It will, in principle, accommodate every type of laser and microwave linkage between platform nodes that are satellite based, airborne, seabased and groundbased, as conceived by the communications architect and designer today and anticipated through the next few decades. In a way the 5-GENIN may also be considered as an implementation of the transformational communication architecture (TCA).

In terms of the airborne platforms considered in the 5-GENIN system, one may include rotary- and fixed-winged aircraft, airships, and remotely piloted vehicles (RPVs) at various altitude profiles. In the satellite domain are space platforms at low, medium, synchronous, and above-synchronous altitudes. The ground-based platforms would encompass both fixed- and mobile-type terminals, including the robotic miniaturized unmanned ground-based mobile (MUGM) force elements. The sea-based platforms would include all types of surface and also subsurface vessels.

In all manner of communication nodes—spatial, atmospheric, ground or sea based, and whether fixed or mobile—the need to connect command and control signals, as well as the platforms involved in the collection of observational data, weather data, and situational awareness information for tactical and strategic needs, requires a large multiplicity of links. These must be secure and often must be very wideband, with the signal data being capable of translation from laser carriers to RF carriers and back. Moreover, because of their small size and weight per MHz (or GHz) and noninterference with other signals, laser beams should be the preferred form of communications.

The 5-GENIN is a worldwide communication system. It combines spatial and atmospheric laser downlinks and uplinks to and from selected ground stations, which are connected to Earth-based Internet systems. In this chapter we start by placing the Internet backbone at a synchronous altitude and proceed to demonstrate, by means of multiple laser antennas and associated transceivers for each of the three nodes of the backbone, the capability of receiving and retransmitting signals from and to all parts of the world.

Second, as described in Chapter 6, in order to downlink and uplink it is necessary to select locations for the OGS that are connected to underground (or overhead) fiber cable and to have at least one of the ground stations located in a clear weather environment. The space-based optical antenna will be controlled to aim at that ground station. This requirement immediately necessitates the integration of weather data sources with those platforms communicating with the Earth-based terminals via lasers. However, when the particular platforms are not connected to ground stations, such as sea-based platforms, it is necessary to transform or shift from the laser bands to RF bands. Although this would result in the need to reduce the signal bandwidth, the RF signal can penetrate the weather and transmit the essence of the required information. For purposes of illustration, this chapter will present an example of a pair of weather satellites whose sensor output is processed at a ground station and then made available to the synchronous backbone nodes.

Chapter 7 concludes with a discussion of a way in which borders and also high-value targets (e.g., nuclear power stations) may be monitored by hovering airborne assets, such as airships, for security purposes. The results of the round-the-clock observations would be transmitted by lasers communications to a control center. Tactical field data, security monitoring, and observation, which provide information to the situational awareness system, is also outlined.

7.2 The Synchronous Laser Backbone

As shown in Figure 7.1, the laser backbone is composed of three satellites, deployed 120° apart relative to the Earth's center.

This configuration can provide extremely large bandwidths, of the order of many terabytes for the links between them. Each of the three

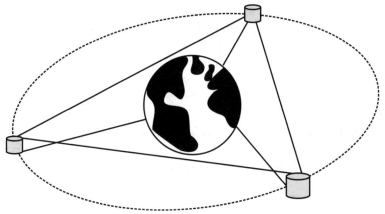

[This diagram courtesy of the U.S. Government]

- Greater channel capacity (through terabytes per second)
- Large number of channels
- Multiple private links to A/C and ground user
- Small optics—allows many users

(a)

Three (3) synchronous satellite at 120° separation
(plus upper and lower deck supporting telescope antennas)

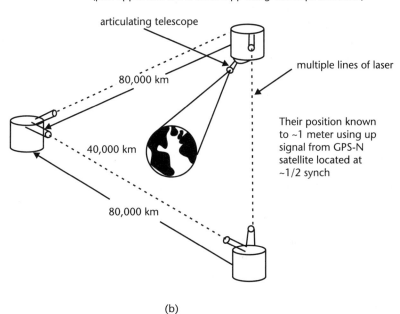

(b)

Figure 7.1 (a, b) Projected laser backbone for the 5-GENIN communication system.

backbone satellites will support a number of optical antennas and associated transceivers, enabling both uplinks and downlinks. This backbone will be the mainstay of the worldwide laser communication system, which could support the Internet function for NASA, DoD, and other governmental, industrial, and commercial organizations. It will also direct point-to-point communications between the various nodes of selected network architectures.

The intersatellite power budget is a concept useful for the design of the synchronous backbone, which was derived in Chapter 2 as (2.7) and is now repeated as (7.1), where n′ is the number of photoelectrons per bit,

$$n' = P_\tau L_\tau G_\tau L_R Q (L_{P-\tau})(L_{P-R}) / L_S h \nu f \qquad (7.1)$$

and the signal power budget for the link between the backbone synchronous satellite and a low-altitude satellite with the range between them is R^1. In other words, the space loss, L_S', is $(4\pi R')^2 / \lambda^2$, making the signal power budget

$$n' = P_\tau L_\tau G_\tau L_R Q (L_{P-\tau})(L_{P-R}) / L'_S h \nu f \qquad (7.2)$$

While the difference between the above two equations is trivial, it is clear that this is the case, because there is no atmosphere or weather to be concerned with and any pointing losses can be removed by zeroing out the jitter and other vibrations affecting the communication links, using the ameliorative techniques described in Chapters 2 and 3. The remaining difference between (7.1) and (7.2) is a function of the distance between the synchronous satellite and high-altitude airborne platforms and the distance between the synchronous satellite and the low-altitude satellite.

7.2.1 Weather Effects

An example of the multiple optical antennas on a single satellite platform, typical of each of the three satellites of the backbone, is drawn in Figure 7.2. An overview of the 5-GENIN system as it interacts with ground-based platforms, with emphasis on weather data that needs to be transmitted to the backbone, is sketched in Figure 7.3.

As we noted in Chapter 6, with the OGS connected to a fiber cable network, it is necessary for one of the telescope antennas onboard one of the satellites of the backbone to be pointing to a selected OGS that has no weather cover above it. The fiber network is the only way to get the laser signal to ground station locations that are covered by foul weather. In this way we are able to connect with just about any location in the United States; and

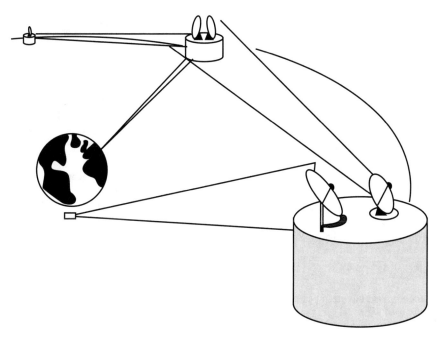

Figure 7.2 Artist's concept of laser backbone links to Earth and atmospheric platforms.

when considering worldwide needs, using the submarine cable and associated connectivity with land-based fiber cable throughout the world would enable us to communicate to any location on earth. Countries such as Peru, Chile, Australia, Kuwait, China, and South Africa represent specific countries that could have optical ground stations, with connectivity established on a point-to-point basis, as well being potential gateways to the remaining countries of the world.

7.3 Example of Weather Satellite for 5-GENIN

In this section we show an example of two geosynchronous operational environmental satellites (GOES) that would be helpful in providing us with useful weather data. This, in turn, will enable us to point the satellite telescope antennas to the desired ground station. The two weather satellites, the GOES-East, which can be deployed, for example, at 75W longitude, and the other, the GOES-West, which would then be deployed at 135W longitude, observe the weather over a large portion of the United States. The processing of the two satellites' sensors may be carried out at a processing center station

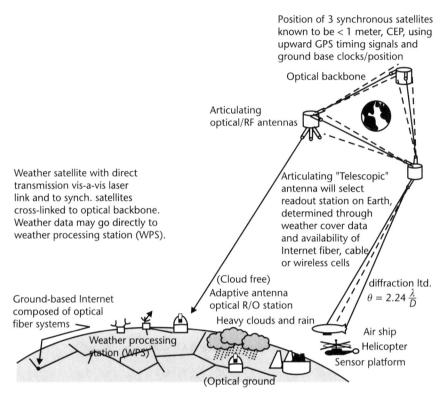

Position of 3 synchronous satellites known to be < 1 meter, CEP, using upward GPS timing signals and ground base clocks/position

Optical backbone

Articulating optical/RF antennas

Weather satellite with direct transmission vis-a-vis laser link and to synch. satellites cross-linked to optical backbone. Weather data may go directly to weather processing station (WPS).

Articulating "Telescopic" antenna will select readout station on Earth, determined through weather cover data and availability of Internet fiber, cable or wireless cells

Ground-based Internet composed of optical fiber systems

(Cloud free) Adaptive antenna optical R/O station

Heavy clouds and rain

Weather processing station (WPS)

diffraction ltd. $\theta = 2.24 \frac{\lambda}{D}$

Air ship

Helicopter

Sensor platform

(Optical ground

Figure 7.3 The 5-GENIN laser communication links with RF backup in space and in atmosphere based on bandwith requirements and weather mitigation.

at Maui, White Sands, or Wallops Island. The overlapping of the two GOES sensor downlinks may be helpful in the signal-processing protocols. Further, in some instances there can be simplification in the connectivity by using a communication relay satellite, at 105W longitude, for example, to combine the 2 GOES satellites' output, for retransmission to the processing station at White Sands. Figure 7.4 shows the intended location of the GOES satellites together with one of the three backbone satellites. The latter also functions as the relay satellite within the weather satellite configuration. [1]

The key benefit of the GOES system is that useful weather data is provided in near real time, which can then be used to direct the laser beam from a relevant backbone satellite to the appropriate ground station. As indicated in Figure 7.4, there are three ground stations (Maui, White Sands, and Wallops Island) that can perform the signal processing of the output of the sensors deployed on the GOES-E and GOES-W. When found necessary, one of the backbone satellites can also be used to relay the environmental data to be processed at an alternate ground station.

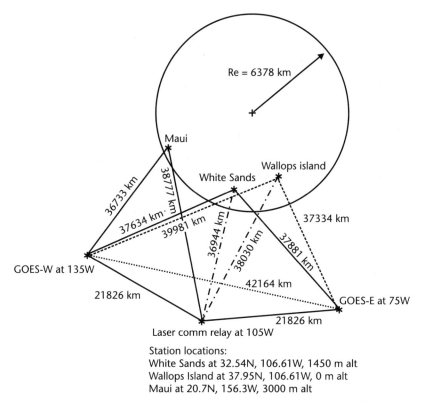

Figure 7.4 Proposed geometries of locations of geosynchronous operational environmental satellites (GOES): GOES-West and GOES-East, with a potential relay satellite as one of the three backbone satellites.

7.4 Unique Requirements of Atmospheric and Earth-Based Laser Terminals

It is necessary that all terminals be equipped with an AO telescope antenna in order to correct for the downlink distortion and then effectively correct the distortion that would typically be imposed on the uplink. As we have seen in earlier chapters, implementation of the AO subsystem should result in considerable reduction in BER.

In terms of locations of the laser antennas in atmospheric platforms; the telescope antennas can be mounted within a bubble on top of the fuselage of aircrafts, on a despun platform deployed on top of the shaft of the rotor of helicopters, and in the bubble located on top of the UAVs.

Similar AO telescope antennas will be included in the design of stationary ground stations and the mobile ground stations and vehicles. An advanced robotic system, known as the MUGM system, will require a small-sized AO telescope antenna system [2]. The proposed MUGMs, operating singly or in a swarm, will be part of the advanced automated battlefield system concept now being developed. The MUGM is particularly useful in tactical situations when there is very limited manpower available, and the required tactical operation is in a lethal and stealthy environment.

7.4.1 Protection of Stationary, High-Value Targets

An example of the use of sensor platforms to observe the performance and monitor the security of a high-value target, such as a nuclear power plant, is shown in Figure 7.5.

What is unique in this configuration, apart from the helicopters and UAVs, is the use of airships. These are designed so that arms extended from the bottom structure would physically support a laser transceiver and telescope antennas, to which a sensor suite output would be connected. The laser uplink signal would be transmitted to the backbone, and from there to a

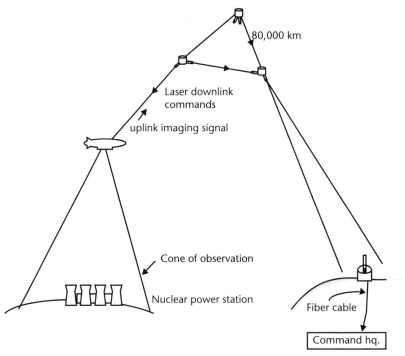

Figure 7.5 Use of airship to hover about high value targets (nuclear power stations, refineries, chemical plants, etc.) for surveillance and monitoring, transmitting images and alarm data to command headquarters, and receiving specific command data.

central command center located in the airborne command post or at a ground-based station in the zone of the interior (ZI).

7.4.2 Observation and Monitoring of Borders

The use of airships to support sensor suites coupled to onboard communication subsystems while hovering over long border areas indicates an approach that would be more cost-effective than the employment of walls and personnel stretched out over several thousands of kilometers. Here again the active and passive sensor data would be transmitted via the laser communication subsystems to the laser backbone at a synchronous position (Figure 7.6). (This type of security monitoring system may also involve battlefield applications.)

A special advantage of the airship is that it can slowly drift at an altitude as high as 12 mi above the border with active and passive sensors that have high range and resolution, enabling it to detect contraband as well as unauthorized people penetrating the border. Also deployed on the airships could be the Joint Biological Point Detection System Suite (JBPOSS) to search the atmosphere and report about any poisonous contaminants in the ground and the atmosphere.

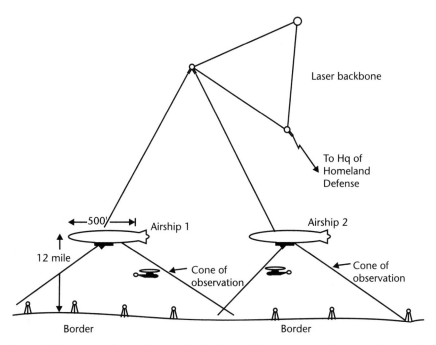

Figure 7.6 Border patroling by airships and helicopters with laser communication via backbone to the command center.

It should be mentioned that the JBPDS was developed by New Jersey Institute of Technology (NJIT) and that the special Disaster Managements Interoperability Service (DMIS), which enables different computers and software to properly interface with one another, is a Department of Homeland Defense program that involves all aspects of border penetration, consisting of biological and chemical releases with a proposed laser communication transmission.

7.4.3   Airship Vulnerability Issues [3–5]

The availability of airships as a support element in tactical and strategic operational situations requires an assessment of their vulnerability. The main concern was always the flammability of the gas that is used in the airship. However, hydrogen has not been used for a long time. It has been supplanted by inert gases such as helium. Importantly, helium gas puts out fire.

In terms of airship survivability, it is similar to the survivability of a C-17 or a 747 airplane, and it may be even less vulnerable because it avoids airfields that are common targets and does not fly in the normal air lanes or commonly used sea lanes. In recent army studies at the Center for Army Analysis, it has been determined that its radar cross section makes it difficult for large missiles to lock on. Further, its slow speed is below the engagement profile of most large missiles, so that it would not interact. And if it did, the fragmentation would only create small holes rather than large ones that would result in dangerous tears. The explosive blast has little effect. As it turns out, a 100-ft bag of helium placed between the explosion and a person nearby offers good protection. Moreover, deflation rates following the creation of perforations are also very slow.

In another form of attack, a heat-seeking missile would more likely hit the engines, which are at the ends of a power-wing the size of a 747 wing. Finally, it is difficult for jets to shoot down airships when using guns, because the size and slow speed make it difficult to aim accurately.

In terms of ice, wind, and lightning, we have the following natural survivability situations: airships can handle extreme cold, snow, and ice while aloft. (There were no blimps lost in World War II due to icing.) In the 1950s, a three-year navy study found icing at the envelope occurring only in freezing rain or freezing drizzle. These conditions occur in the Northeast portion of the United States in three to five storms per year and do not extend more than 200 mi out to sea. In other examples, in 2004, icing occurred in France 3% of the time. In warmer waters such as in the Pacific Northwest, icing conditions are much less frequent.

While icing and snow at the mast are dangerous, new materials are now being used which resist "sticking," so that ice and snow do not accumulate very much. Newer designs incorporate anti-icing capability; these are similar to the material now used in fixed-wing aircraft. It should be pointed out that the Goodyear GZ-22 airship builds up ice only at the nose, the strongest part of the airship, at speeds of 50 kn. For the ice at the nose of the CL-160 airship, the speed has to be 52 kn before buildup occurs.

In terms of the effect of wind, while the airship is on the ground, the threat is greatest, in that the airship may tear away from its mooring and hit another structure, resulting in a large tear and followed by serious deflation. However, in sample strategic deployment routes of the airship, there was little effect of wind at the altitude the CL-160 airship uses, from a 1-kn headwind to a 5-kn tailwind.

In a lightning environment, the effect is essentially nonexistent, since the airships are built as Faraday Cages. Thus, the lightning passes through the helium and the envelope of the airship, which are both nonconductive. Moreover, the static charge is dissipated by diverters, just as in fixed-wing aircraft. Historically, airships such as the Graf Zeppelin were often hit by lightning without any significant effect.

7.4.4   Endurance of Aerostat After Suffering Enemy Fire

In terms of Aerostat endurance after damage, based on its 590,000 ft^3, which is roughly 32 times smaller than Airship CL-160, if 25 rounds using 50-caliber machine gun fire hit the aerostat and produced 50 holes, it would continue to fly for two hours at its altitude of 15,000 ft and then descend during the third hour down to 1,000 ft and then to the ground, but it would remain able to fight during the three hours.

When hit by a Stinger or a folding fin artillery rocket, the Aerostat would stay at its 15,000-ft altitude for about 45 min and then descend to 10,000 ft in the next 30 min, and then down to earth. The total time from being hit to the ground position is 1.5 hr, while its mission capability time is 1.25 hr.

When enemy action consists of 50 rounds from a 20-mm cannon, producing 100 holes, the Aerostat will stay at its 15,000-ft altitude for about 25 min and descend to 10,000 ft in the next ~10 min. The total mission capability is estimated to be 36 min.

It should be emphasized, however, that the application of the Aerostats and airships at this planning time involves the use of these dirigibles for border patrols and monitoring of any unauthorized penetration of the boundaries of high-value targets. In association with UAVs and/or helicopters protecting the lighter-than-air vehicles, carrying sensor suites can be sufficient, especially when the dirigibles are operating along and within the borders of interest. Application of the particular version of the airship design for use in tactical military field engagements in battlefield areas may also hold benefits.

7.5   The Miniaturized Unmanned Ground-Based Mobile Systems

Apart from the space-based, airborne, sea-based, and ground-based observational and force-application subsystems, we introduce here a key element in the array of the projected automated battlefield components: The MUGM system (Figure 7.7).

The MUGM is composed of the ground-based dual of the UAV. Its elements include automatons (very small tanks and other unmanned special-purpose vehicles), MUGMs, and microminiaturized mobile systems for ease of penetration of enemy lines. The use of laser-based transceivers on space and terrestrial platforms will also be helpful to monitor and control the MUGM in the battlefield, against enemy personnel and their manned mobile and stationary equipment and facilities. The utilization of the MUGMs should dramatically reduce the number of our troops required in any combat mission.

Depending on the tactical mission, the MUGM unit, weighing 10–12 lbs, would contain a sensor and communication system and sufficient power to be directed to observe and target enemy resources in house-to-house fighting. The sensor suite's data is transmitted to a UAV or a manned platform, or

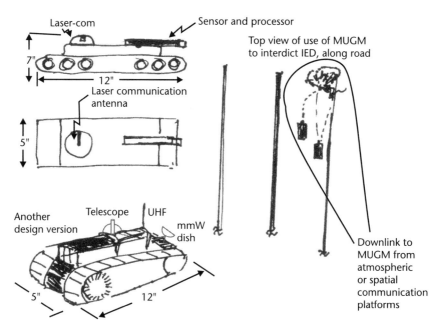

Figure 7.7   The miniaturized ground-based unmanned mobile elements, singly or in a group, search for and detonate an improvised explosive device.

to a low-altitude satellite directly to the backbone, for retransmission to the command headquarters. Other MUGM elements may be loaded with high explosives and be directed by a laser link to travel to a specified target and take it out. This order will be implemented by either shooting a small missile or by the MUGM exploding itself when reaching a specified distance from a target.

7.5.1 Unique Applications of the MUGMs [2]

Because of their unmanned small-sized structures and ability to be moved autonomously, the MUGMs can be directed by a laser signal to search the sides of the road for any mines and other lethal devices, such as the shaped-explosive devices that can typically hurt our vehicular patrols, convoys, or civilians in general. By having the MUGMs do the patrolling, they would be able to take out such devices and save soldiers' lives. The MUGMs can also be trained by way of their pattern recognition subsystem, which is integrated to its day/night video sensors, to recognize the explosive devices. Those would be destroyed by the MUGM by acoustic means or by shooting an explosive ordnance.

7.6 Ground-Based Power Support for the Backbone Satellites

A support technology for the three synchronous satellites in the backbone configuration could provide a laser-based power beam from the ground to focus uplink energy into the satellites' photovoltaic arrays. This appears to be a useful and efficient system engineering concept, in that the current design of energy storage, such as batteries, is a large portion of the mass of the satellite power system [5].

In a similar way, the ground-based laser energy-delivery system may substitute for most of the mass of the lunar power system, which would be used in supporting a laser or an RF communication system, for transmitting sensor data to the Earth. While NASA's Lunar Laser Communication System has been canceled for the next decade, there will be an RF communication system, likely at the K_a band. That system would benefit from energy transmission support from the Earth.

Since solar cells are more efficient under monochromatic radiation than the broadband solar radiation, the use of the ground-based laser, because of its narrowband capability and having its beam directed, with the aid of an adaptive optical subsystem, to the appropriate satellite in the backbone configuration, makes the overall power system system efficient. During an eclipse, which lasts for a short period (70 min or about 5% of the orbit), no added subsystems are needed for the satellite, as the solar array needed to receive the power is already in place. With laser power required for just 70 min per day, or 90 days of the year, this does not adversely impact the satellite design. In fact, the time spent on the ground allows for repair and adjustment of the laser system, thereby enhancing the operational performance of its power delivery. It should also be noted that the same ground-based laser has the robustness to provide energy to other synchronous and low-altitude satellites, which are not part of the backbone.

A study by Landis [6] further expands the advantage of the use of a ground-based laser to direct energy to the satellite. It indicates that extending the satellite life after the batteries have died would save more $100 million per year for a satellite weighing several hundred pounds. This further enhances the ground-based laser concept and its importance in delivering power to the satellite.

There may be a question of whether a ground-based RF power delivery system may be considered as an alternate to the ground-based laser. First, a high mountaintop in a dry region needs to be chosen for the RF power station. And to make it competitive, particularly because of its antenna size, a frequency of 94 GHz (wavelength of 3.2 mm) is needed. By comparison, the laser wavelength would be 0.84μ with a transmitter system of a free electron laser (FEL) and an AOS with a 12-m aperture, while the RF will require a phased array 1 km in diameter. The overall efficiency of the ground-based laser is estimated to be 2%, while the RF's efficieny is 0.05%. Moreover, the existing satellite's photovoltaic array will be used for the absorption of the laser pulses, but in the RF case a special antenna would be required.

7.6.1   Additional Features of the 5-GENIN System

The 5-GENIN communication system involves the transmission of all communication links to the backbone for retransmission to intended callees, which could be via a central office or a command center that is collecting observational data and issuing command messages. Moreover, to phase in satellite communications employing RF bands, there would be the need onboard the backbone satellites' antenna subsystems to receive and transmit both RF and laser bands. Such an antenna is shown in Figure 7.8.

Thus, close to the triangular deployment of the laser backbone there would be the triangular deployment of RF satellites in a backbone configuration (Figure 7.9). The distance between each of the laser satellites and the RF satellites would be of the order of several thousand km and up to 10K km. The RF carriers would likely be in the Ka-band, while the laser wavelength would be in the range of 0.5–1.5μ.

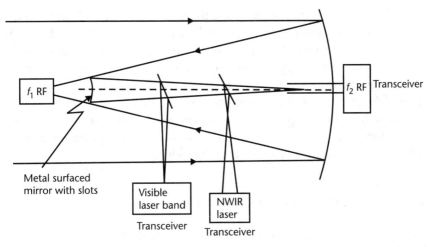

Figure 7.8 RF/laser Cassegrain antenna with integrated transceivers in two separate RF bands and two separate laser bands, in the visible and near-wave infrared.

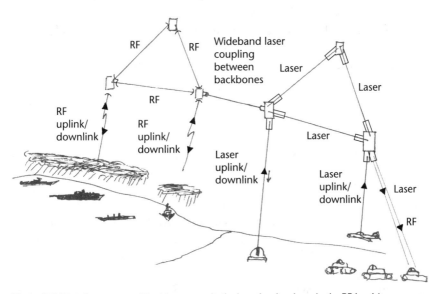

Figure 7.9 Two interconnected backbones, one in the laser band and one in the RF band (to accommodate existing RF communication equipment onboard surface ships). The tanks may have combined laser and RF communication equipment. (Both the laser and RF synchronous backbones circle the Earth.)

As an example from ships at sea, the uplink would be RF going to the RF-backbone and from there to a downlink RF, and also in parallel be converted to a laser signal and going down to a ground-based headquarters in the United States. The atmospheric platforms would include airplanes of various fixed-wing and rotary-wing types; UAVs of all types; airships and aerostats; ground-based stations, both fixed and mobile; tactical equipment such as tanks and MUGMs, ships at sea, and underwater craft. However, the RF and laser communication systems would be combined, starting with the existing RF allocations and moving upward to the wideband laser bands.

7.7 Summary and Concluding Remarks

The major contribution of this chapter is the introduction of the 5-GENIN backbone system. It is composed of three wideband laser communications satellites with cross-links and uplink/downlink capability. The three satellite platforms are deployed at a synchronous altitude in a Y pattern, with each satellite separated 120° from each other and the center of the Y at the center of the Earth. Uplink and downlink communications from the synchronous backbone to the Earth stations may be fixed or mobile, using rotary- and fixed-wing aircraft, airships, and ground-based reconnaissance and force elements such as the MUGM system. The laser links from these elements are fed to the synchronous backbone and from there to their intended callee terminals.

Within the synchronous backbone platforms there will be the associated encryptors/decryptors, servers, processors, memory systems, routers, and adaptive optical antennas, as required to handle messages from and to laser terminals around the world. Emphasis is placed on laser communications to and from airships, because of the latter's ability to hover about borders and high-value target facilities, which require monitoring and response. Also of value is command and control laser communications (CCLC) between a command center and the MUGMs force and reconnaissance elements, which are important in conflicts involving terrorism. The wideband private laser links enable cameras based on the MUGMs to take high resolution photos of the IED, mines, and other explosive implements the enemy may use. A command signal issued to an RPV or a MUGM will target those devices.

The 5-GENIN system will enable a combination of RF and wideband laser communication to be supported from each Earth (ground and ocean)- and atmospheric-based platform. The call from any of the latter stations and terminals would be repeated at the synchronous backbone, for retransmission to specific callees in various parts of the Earth. In the next twenty years, it is likely that the bandwidth requirements will become large enough to benefit from the processor equipment onboard the synchronous backbone.

References

[1]   Field, R. Briefing titled "Optical Crosslink Option for Next-generation GOES," July 2003.

[2]   Aviv, D. G., "The MUGM System," Presentations at the RAND Corporation, July 1992, July 1998.

[3]   Woodgerd, M., Center for Army Analysis, June 2001.

[4]   Cesar, E. M., and Noah Schlachman, "Heavy Lift Airship for Stryker Brigades," *DARPA Watch*, 2004.

[5]   Bolkom, C., "Potential Military Use of Airships and Aerostats," National Defense and Foreign Affairs, Defense and Trade Division, Order Code RS 21886, Congressional Research Service, July 15, 2004.

[6]   Landis, G. A., "Laser Beam Power Satellite Demonstration Application," Sverdrup Technology, NASA Glenn Research Center, Cleveland OH 44135, and L. H. Westerlund, Satellite Communication Consultants, Rockville, MD 20853.

8

Passive Reflector Configurations

8.1 Introduction

In this chapter, a review of the Relay Mirror Experiment (RME) [1], which created opportunities for a variety of laser propagation experiments as well as the demonstration of the utility of reflector mirrors in space, is presented. The experiments, performed principally by Ball Aerospace and Systems Group (BASG) of Boulder, Colorado, and its team under the management by the Strategic Defense Initiative Organization (SDIO), provided a number of useful results, thereby enabling the implementation of a number of system concepts described in this textbook.

A major aspect of the RME in terms of the use of mirrors on a spatial platform is highlighted in this chapter. The articulating mirror system (AMS) and also a nuanced version of the RF Westford experiment, the Optical Westford System (OWS), are presented. The AMS may lead to a number of different applications in communications, ladar (laser radar), and energy deposition, while the OWS may perform relatively short duration, second-order communications functions.

The RME, because of its success in the use of mirrors in communication missions, validated the design approach of an ordinary flat mirror and also the retroreflective mirror of tubular design, which can be used for protection and for specialized applications. The selected design of the passive reflector system for communications also implies its potential use in active and passive sensors and in energy deposition. These are all derivable from the basic RME demonstrations [1, 2].

There are a number of advantages in considering passive reflector structures. The primary one is that of being able to extend the propagation distance of the laser beam toward a receiver platform, without an active laser relay subsystem and its required power supply. In Figures 8.1 and 8.2, the basic propagation links are shown, presenting the spatial reflective structure with the Primary Signal Source on the ground, followed by a variation, in which the Primary Signal Source is in space, a distance away from the passive reflector [3].

If the selected reflector structures in space are controlled mirrors (Figure 8.3) whose supporting platforms are station-kept, it will enable the laser, whatever its origin, to be pointed, by reflection, to any spot in space, ground, sea, or an atmospheric-borne platform. Those may include fixed- or rotary-winged aircraft, a UAV; on the ocean, any surface and underwater vessel; and on the ground, any stationary terminal, moving ground terminal, and MUGM force elements [4], provided those platforms' optics are within the available field of regard of the articulating mirror. However, there are cases when the combination of desired geometrical line of sight between the reflective mirror elements and the intended receiver optics may be partly obscured by the supporting structures. In fact, the optical beam might be occluded by the timed orbital location and geographical location of the Earth's platform terminals. These conditions will clearly tend to reduce the performance of the communication link.

Actually, in any realistic reflector design configuration, there will be a reduction in signal intensity, and, in consequence, a reduction of the data rate capacity of the reflected and received beams. This can be demonstrated by calculations, and it is due to the fact that only a portion of the directed signal from the actual laser communication source is intercepted by the deployed mirror. Then, only a portion of the reflected signal photons aimed to the intended receiver is collected by the receiver optics. But by enlarging the mirror diameter, narrowing the beamwidth of the primary source and the reflecting subsytem, the relay configuration will nearly be able to meet the bandwidth requirements. While in the RME experiment the primary signal source was uplinked from a ground-based laser transmitter, and the reflected signal was downlinked to a ground station, the primary signal could very well be located in space, astride the spatial mirror, and reflected to the ground station. Both approach paths are discussed next, but in terms of the various reflective structures, only two are selected. Those reflecting structures are the articulating mirror and the optical dual system of the RF Westford Experiment. The latter, composed of a very large number of tiny mirrors, is called the Optical Westford Experiment.

In terms of the mirror, whether a flat plane, cylindrical, small or large, configured in a retroreflector mode or a straight reflector is considered, its

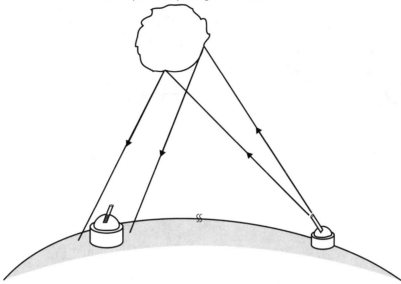

Arbitrary reflective surface enabling
uplink/downlink communications
between separated optical ground stations

Figure 8.1 Laser beam emanating from a ground station (primary signal source) toward an arbitrary reflective surface structure in space and from there redirecting a beam via downlink to another optical ground station.

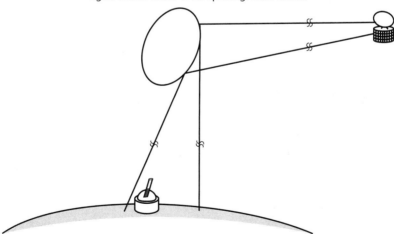

Arbitrary reflective surface enabling
the transmission from a space-based primary
signal source down to an optical ground station

Figure 8.2 An arbitrary reflective structure in space, reflecting the beam from a primary signal source located in space a distance away from the reflector, which directs the laser beam to an optical ground station.

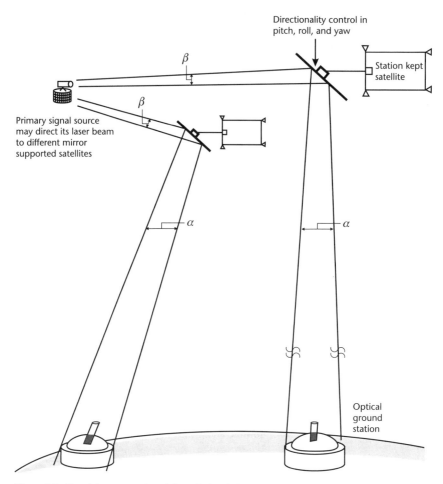

Figure 8.3 Pictorial representation of the articulated mirror satellite.

application in communications is most efficacious. Its usefulness is also extendable to ladar (laser radar) and to direct energy (optical or infrared bands) to an intended target such as described in the RME for possible use in the Space Defense Initiative (SDI) domain [2, 7]. Such applications are emphasized in this chapter and also in Chapter 9.

While a mirror is costlier for low-data-rate applications, in the wide bandwidth reflection genre, it will require a supporting satellite, which will increase the system cost but keep the cost per MHz low. Typically, such satellites would weigh about 200 lb (~90 kg); this satellite will be overviewed later in this chapter. However, because of advances of composites and various nanotechnology components, smaller satellites of lower weight, which could be used to support the mirrors, are likely to weigh less than 50 lbs (~23 kg).

In the communications field, the reflected laser beam might be useful in a number of tactical applications, including secure communications, spoofing, and other functions. And it might even weigh less than 10 lb (~4.5 kg). Conceptual designs composed of an array of a phased combination of microplatforms can become a supporting platform for the articulating mirror. Apart from tactical communications, which can beneficially utilize the low-weight support structures, reflection of high-intensity laser beams from a hardened mirror that is supported by microsatellite platforms might also be feasible in future system design.

In terms of actual flight experiments employing a mirror reflector, Ball Aerospace and Systems Group flew the RME spacecraft (S/C). It demonstrated the ability to acquire, track, and point, and also to control a laser beam transmitted from the Earth and reflected off its bottom floor and back toward a target on the Earth. Because of the use of mirror to reflect visible and near-infrared laser beams, which have a variety of mission applications, it is discussed at length in this chapter [4].

In the section to follow, a brief history of the passive reflector satellites for the RF is presented, followed by its optical dual. It then concludes with a brief description of a potential reflector surface that will enable the reflection of a combined RF and optical beams.

8.1.1   History of Passive Reflectors, the RF Case [5, 6]

In the early days of satellite communications, passive reflectors rather than active repeaters were deployed on the satellite platforms. Following the experimental transmission of a narrowband Morse Code signal directed at the near full Moon and getting the reflected signal message back at a receiver station thousands of miles away from the transmitter, a more effective passive satellite experiment was considered: Project Echo. This was in order to achieve a wider bandwidth signal. It consisted of launching two spherical structures as the reflector bodies. One was launched in 1968 and the second in 1969. The first satellite was a 100-ft-diameter sphere, and the second, a 135-ft-diameter sphere. The orbit of the *ECHO 1* was 820 nm · 911 nm, with a 48.6° inclination, and that of *ECHO 2* was 557 nm · 710 nm, with an 85.5° inclination. However, neither of the spheres was station kept.

The spherical antenna structure reflecting a laser communication beam might have a number of attributes, particularly when the platform is station kept, but in the RF-signal case, the reflective *ECHO 1* and *ECHO 2* had their design emphasis placed on the reflectivity of the impinging RF signals, without any amplification or control electronics [3]. The frequencies used in *ECHO 1* were 960 Mhz and 2,390 Mhz; 160 Mhz was used with *ECHO 2*. The spherical cover material of the satellite was aluminized mylar. Signals were transmitted between concurrent mutually observed stations in the United States and the United Kingdom. Useful tracking data was also gathered from *ECHO 1,* including radar cross-section measurements. However, there was a large surface-diameter-to-weight ratio; for example, the surface/weight of *ECHO 1* was 100 ft/166 lbs (~30.48m/ 75 kg), and that of *ECHO 2* was 135 ft/547 lbs (~41m/ 249 kg), causing the structures to suffer from solar interaction perturbations, leading to high drift rates [3].

Later in this chapter, the mirror supported structures will be reexamined, but with controlled stabilization, for optical laser signal reflectivity for communications applications. Ladar applications using passive mirror structures will also be outlined in Chapter 9, which covers a variety of laser applications. Ladar is included since the return signal often requires its communication to a command center.

The high-energy reflectivity via mirrors is also outlined, since there is considerable interest in its implementation in the field of SDI [1]. However, by linear extension, if the spatially supported mirror can be used to relay high-energy laser pulses, clearly it will be capable of reflecting low-energy pulses, which are commonly used in laser communications. In RME, the reverse logic is deducible.

8.1.2   Extension of Applications in the Passive Field

Before we get into the details of the optical communication applications using reflective mirrors, it is important to mention a unique experiment in the RF band, in which the RF reflective structure was a bundle of a large number of thin needles. Each was the length of a dipole; that is, half the wavelength of the signal carrier that is intended to be reflected. The experiment, called Westford and developed by the MIT Lincoln Laboratory, consisted of dispensed little dipoles, at an altitude of about 2000 nm (~3700 km) in an orbit that is polar circular. Each of the 480-million dispensed copper needle dipoles was 0.72 inches (~1.829 cm) long (carrier was 8.350 Ghz) and 0.0007 inches (~0.00178 cm) in diameter. The weight of the 480-million needles was about 43 pounds (~19.55 kg), and it was successfully launched onboard an Atlas-Agena B vehicle. The launch took place in May 1963. Both voice and 20 Kbps of a frequency shift keying (FSK) modulated data signal was transmitted from Camp Parks, CA, to a ground station in Westford, MA.

The initially concentrated bundle of dipoles enabled the 20-KBps signal to achieve an adequate S/N and be reflectively propagated forward. However, as the dipole cloud started drifting only a small portion of the original cloud was left to interact with the upward signal, and the reflected signal became very weak.

The launching of a low-altitude spacecraft with the bag of dipoles that would be commanded to open and release the dipoles, was the specified experiment at that time. Since then, three different variations of the demonstrated principle and application of this modality have been tried. First, today's component and subsystems technology makes the electronics onboard the small spatial platform simpler, more reliable, lighter in weight, and easier to launch and deploy payload packages. The success of the recent microsatellite platforms, the XSS-11, for example, singly and potentially in groups, is likely to make that configuration useful for wireless service, particularly in utilizing intermicrosatellite laser communications.

The second application of releasing particulates in space, considered principally by Harris Mayer of the Aerospace Corporation and Stan Sadin of NASA, was to release fresh ozone gas into various regions of altitude and latitude, where the ozone layer gets depleted. An alternate method was to release a set of chemical compounds which, under the UV radiation from the sun, would produce fresh ozone to replenish the depleted layers.

The third application was suggested by Edward Teller while at the Hoover Institution. It is based on the fact that when a large amount of hot ashes and lava were spewed out during the eruption of the Mount St. Helens, a large number of particles at high altitude including ash-dust clouds were found hovering for months over certain parts of the world. These caused the partial attenuation of the Sun's rays, resulting in a temperature decrease on the surface of the earth of $6°$. He recommended that NASA initiate experiments whose goal is to develop methods of using ashlike substances to reduce the Earth's warming. The frequency of release of the particulates, their chemical composition and disintegration rate, as well as the frequency of launch were the topics of his study.[1]

8.2 The Nominal Reference Link

Establishing a nominal reference link for the reflective structures is necessary in order to evaluate and compare the performance of the two different passive structures discussed here. The calculation of the nominal reference link is carried out using a specific data rate and modulation scheme, and the distance between the passive structure and the ground station. Then, for each of the two reflective structures, the resulting communication performance, relative to the nominal reference link, is determined. As will be seen, the key factors

---

[1]This author had the honor of suggesting these approaches to Carl Sagan of Cornell University, who was keenly interested in the proposed experiments.

in these evaluations are the reduction of number of photons per bit and the associated reduction in data rate, from the nominal reference link.

8.2.1   Data Rate, Modulation Scheme, and Range

The nominal reference data rate is taken to be 1.0 Gb/sec, with PGBM and synchronous distance from the reflector structure to the ground station of 40,000 km. An evaluation of the resulting data rate for that same geometrical position, for different mirror platforms, will then be undertaken.

Repeating the signal power budget expression, (2.6), derived in Chapter 2, for n′, the number of photoelectrons per bit, we have:

$$n' = \left(P_T G_T L_T L_P L_A L_R G_R Q\right) / \left(L_S f h \nu\right) \tag{8.1}$$

where

P_T = optical power in watts
G = gain of transmitting optical telescope = $(\pi Dt / \lambda)^2$
L_T = losses in transmitter system
L_P = pointing loss
L_A = losses in atmosphere due to turbulence and weather
L_R = losses in receiver system
G_R = gain of the receiving optical telescope = $(\pi Dr / \lambda)^2$
L_S = space loss = $(4\pi R / \lambda)^2$
R = range between spatial platform and the OGS
$h\nu$ = energy per photon (Joules per photon)
h = Planck's constant = $6.625 \cdot 10 (-34)$ Joules per hertz per photon
ν = frequency of the photon in hertz (cycles per second) = c/λ
f = signal's data rate in bits per second or in hertz

For simplification purposes, (8.1) is assumed to describe a signal that is optically aligned and locked between the reflector, at a synchronous altitude, and the optical ground station, which is assumed to have no interfering weather and no atmospheric turbulences. In practice, however, the atmosphere does indeed interact with the signal photons, and if the reflector platform is not station kept, the structure will certainly drift. Thus, the SPB given previously would only hold for a relatively short periods of time (unless the methodologies discussed in Chapter 6 are in place). However, for the purpose of evaluation and comparison of the different mirror-based structures, the deleterious atmospheric problems are essentially neglected, and unless otherwise noted, the structures are assumed to be station kept.

A numerical example employing (8.1) is presented below to determine the required number of photoelectrons per bit, together with the margin for our nominal reference link. This link is similar to the one presented in Chapter 2.

P_T = 370 mW; –4.3 dBW
L_T = Transmitter loss, –2.0 dB
G_T = for 5.5-inch optics, 116.5 dB
P_R = Receiver Power, –60.4 dB
G_R = for 20-inch optics, 133 dB
L_R = Receiver Loss, –2 dB
L_P = Pointing Loss, 0.3 dB
L_A = Atm-turbulence loss, –2.0 dB
Received energy per bit = 150.4 dBJ
Received photons per bit = 33.9 dB
n′, number of photoelectrons (pe) per bit 29.9 db (~1000 pe)
f = data rate, 1.0 GBPS, 90 dB
L_S = 299.2 dB
hv = energy per photons, 184.3 dBJ
Q = quantum efficiency, 40%, –4 dB

From Figure 2.6 of Chapter 2, we have, when using PGBM with an extinction ratio of 100, 1 photoelectron of noise background and a BER of 10 (–7) errors per bit. The number of required signal photoelectrons per bit would then be 38, or 15.8 dB. Thus, the margin attained for the nominal reference link is 29.9 – 15.8 = 14.1 dB, or 962 pe/bit (photoelectrons per bit). However, as the common value of margin for typical communication links is 6 dB, then 8.1 dB can be assumed to be allocated to atmospheric losses. Alternatively, when we use 18 pe/bit as the background noise, the number of required signal pe/bit would then be 75 and the margin would be 1000 – 75 = 925 pe/bit, or 11.25 dB.

8.2.2 A Simplified Signal Power Budget for a Reflective Structure

The results obtained in Section 8.2.1 may be obtained in a simpler way: multiplying the number of photons per second transmitted by the laser, by the ratio of the receiver aperture on the ground to the entire beam's cross section hitting the ground area, around the ground terminal.

$$n = (P_T / hv) \left\{ \pi \left(d^2 \right) / 4 \right\} / \left\{ \pi / 4 \left(R^2 \right) \left(\theta^2 \right) \right\} \qquad (8.2)$$
$$= (P_T / hv) \left(d^2 \right) / \left(R^2 \right) \left(\theta^2 \right)$$

where

n = number of photons per bit

θ = beamwidth of laser downlink from the reflecting structure, 5 μradi

d = diameter of the aperture of the OGS, 50 cm

R = 40,000 km

P_T= laser transmitter output power, 1W

Substituting these numerical values into (8.2) gives n = 1.285 · 10^{13} photons/sec. And dividing n by the loss factors of $L_T L_P L_A L_R$ = 4.37, we get n = 2.94 · 10^{12} photons/sec.

As indicated, when using the SPB, the number of photons per bit becomes 33.9 dB (2,450 photons per bit), and for the reference data rate of 10^9 bits per second, the number of photons per second becomes 2.45 · 10^{12}.

8.3 Selected Passive Reflectors

Based on the calculations in Section 8.2.2 we are able to compare the two reflective structures listed below and to provide a general description of the evaluation approach, and indicate the key design considerations of the space-based mirror and the Optical Westford Experiment.

8.3.1 The Articulating Mirror System

Shown in Figure 8.3 is a mirror reflecting a laser beam from the primary orbital source to an Earth-based terminal. By virtue of station keeping and the attitude control system (ACS) built into the satellite supporting the mirror, significant tracking capability can be achieved by the mirror movements. The terminal communicating with the primary signal source may be an airborne terminal or a moving ground terminal. The latter may also be a ground-based force element, such as the MUGM.

Based on the conceptual simplicity of the design of the entire reflector platform configuration, it is evident that additional gimbaled mirror subsystems may be appended to the satellite configuration and controlled by it (Figure 8.4). This enables the spatial node to handle more than one signal source for reflective retransmission of the primary signals, to selected Earth stations and/or a combination of stationary and moving ground terminals and airborne terminals.

8.3.2 Calculations of Data Rate for Articulating Mirror System

To perform elementary calculations of the link performance for the AMS reflector, Figure 8.5 is employed.

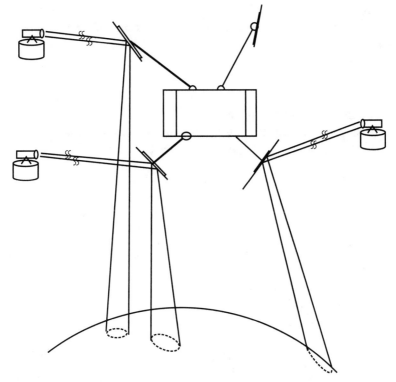

Figure 8.4 Multiple mirror configuration providing separate mirror extensions from a single platform.

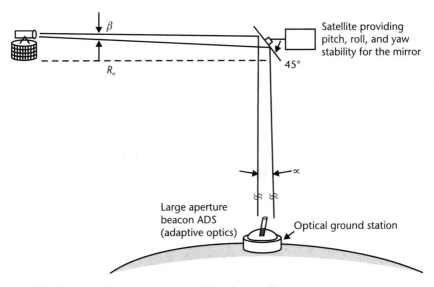

Figure 8.5 Geometrical description of the articulating mirror satellite.

As before, beamwidth, β, of the primary source and Rn, its distance to the mirror, are considered. Consequently, the rectangular size of the mirror flat may be written as

$$\left(Rn\beta\sqrt{2}\right)\cdot(Rn\beta) \tag{8.3}$$

when assuming the mirror tilt to be 45°.

The reference Rg is taken to be 480 nm (889 km).

The power loss term, La, may be expressed as before

$$La = \alpha\gamma/\beta \tag{8.4}$$

where

α = beamwidth of reflected beam along an axis of its cross-section

β = beamwidth of the laser signal from primary satellite optics, assumed diffraction limited, 2.24 λ/D

γ = beamwidth of reflected beam along the axis perpendicular to that of α

For the pointing subsystem, typical numerical values for α and γ that are of the order of 100 μrad with La = 400, would in turn reduce the data rate from the reference link to 2.5 MB/sec. However, a more sophisticated mirror-pointing system at the reflector structure and aiming at the ground station that has a beacon, a large aperture, and an adaptive optical subsystem at the receiving end will enable the data rate to be up to the reference data rate of 1 Gg/sec.

Additional link improvements can be considered apart from the more sophisticated pointing subsystem. As before, these would include an increase in power, decrease in beamwidth, and reduction in margin. A trade table showing these numerical values is given in Table 8.1.

The mirror size in the calculations given in Table 8.1 is based on (8.4) using Rn = 480 nm (889 km), with β as the variable. Clearly a smaller mirror size will result in a lower data rate. This reduction may be calculated—the data rate will be reduced by the ratio of the chosen mirror size divided the maximum size for which 1 Gb/sec was attained.

Because of the importance of the articulating mirror system (AMS), special attention is given here by way of an example of a typical small satellite supporting the mirror and its onboard subsystems (Figure 8.6). The weight of the satellite is estimated to be about 200 lb (~90 kg). But as mentioned before, with composite materials and components using nanotechnology, the

Table 8.1

Articulating Mirror System, Data Rate as a Function of Power, Beamwidth, and Margin for Rn - 889 km

Beamwidth	Data Rate	Laser Power	Margin
5 μrad	2.5 Mbps	0.370W	15 Db
5 μrad	250 Mbps	1W	0 Db
5 μrad	40 Mbps	1W	7 Db
1 μrad	1 Gbps	1W	7 Db
2.5 μrad	1 Gbps	1W	1 Db

weight of the platform could be reduced to less than 50 pounds (~23 kg) and potentially to less than 20 lbs (~9 kg).

8.4 Experiments Using Passive Spatial Reflectors

A multiyear program led by Ball Aerospace & Systems, the RME, was carried out to measure the effectiveness and performance of the spacecraft and associated systems under different operational requirements, including communications links through the atmosphere under quiescent and turbulent conditions, as well as a multitude of considerations based on augmented acquisition and tracking for ground-based laser illumination. In the main, however, the RME measurements were to determine whether energy can be accurately deposited at a particular location on the ground, and later, by extension, on a moving target on the surface of the Earth, or one residing or moving in the atmosphere or space. When the experiment took place in February 1990, RME (Figure 8.7) was one element of SDI performance testing, which was to validate the technology of acquisition, pointing, and tracking for the ballistic missile defense system applications. RME spacecraft (S/C) employed a relay mirror to reflect a 1.064-μ laser, propagating a narrow beamwidth signal from a ground station to the orbiting S/C at 450-km altitude, which was then reflected from the S/C lower deck mirror to a ground target. The total travel of the beam was ~1200 km.

The source of the Nd:YAG laser was at the laser source site (LSS) located at the AF Maui Optical Station (AMOS), at an elevation of 3,000m above sea level. The additional visible laser source uplink concentrically with the NWIR beam, but going only one way, was the 0.448 mm (Argon) laser; it is considered the tracking beacon. The AMOS beam director/tracker provides course

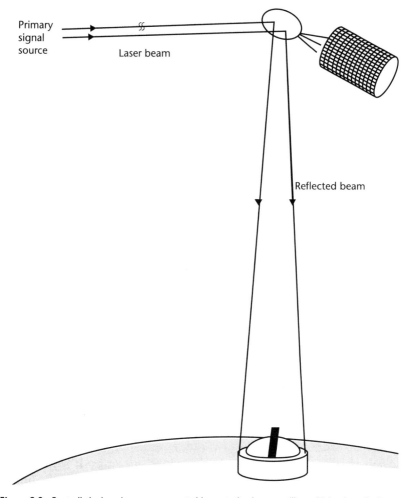

Primary signal source

Laser beam

Reflected beam

Figure 8.6 Controlled mirror in space, supported by a station-kept satellite, which relays the laser communication beam to an optical ground station.

pointing and tracking for the entire system. Two sensors are used to measure the intensity of the two laser beams returning from the S/C's retroreflectors. The output of the two sensors is connected to the augmented tracking and acquisition system (ATAS) control electronics. The output from either sensor is used to control the point-ahead mirror for the beam point ahead and centering on the spacecraft.

The TSS's beacon laser is generated on the optical bench and directed to the gimbal assembly. This gimbal provides the coarse beam pointing and tracking for the system. The beam is then directed out of the gimbal through

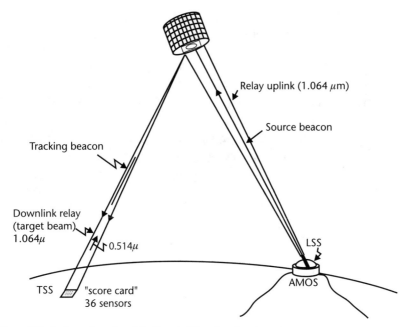

Relay uplink (1.064 μm)

Source beacon

Tracking beacon

Downlink relay
(target beam)
1.064μ

0.514μ

LSS

TSS   "score card"
36 sensors

AMOS

Figure 8.7 Basic configuration of the Remote Mirror Experiment.

the center of the target board. Thirty-six NWIR sensors are distributed over
the target board. Applied to each of the 36 sensors is a small telescope. Additionally, there is a wide FOV camera, which is used for initial acquisition of
the S/C and is attached to the perimeter of the target board. The output of
this sensor goes to the ATAS electronics and the point-ahead mirror.

Because of the need to measure other aspects of the SDI realm, a large
number of different subsystems were deployed vis-à-vis the two ground-
based laser stations and the satellite platform, which, among the onboard
instruments associated with demonstration of ATP and vibration assessment
and servo-based amelioration, included the ordinary mirror and retroreflector mirror. The mirror subsystems validated by implication the design of the
types of mirrors discussed in this text.

Ball Aerospace Systems Group was responsible for the design and the fab-
rication of the S/C and the payload experimental package (PEP), the ground-
based laser sites, and mission operation. Members of the RME team also
included Applied Technology Associates, AVCO/Textron, Bendix Field Engineering, Rome Air Development Center (RADC), and Sunnyvale Satellite
Tracking Center (SSTC). The entire program, however, was managed by the
Phillips Laboratory for the Strategic Defense Initiative Organization (SDIO).
The reports and papers addressing the RME and measurement program are

abundantly available through *SPIE* Volume 1482. Figure 8.7 shows the essence of the RME configuration [1, 2] by way of a cartoon.

The most relevant portion of the RME to our laser communication text is having an uplink to the spacecraft's mirror structure, composed of three laser lines and going from one ground station, known as the LSS, to another ground station, the TSS. As noted earlier, of the laser family, two are argon ion, at 0.488 µm and 0.515 µm, and the third is an Nd:YAG: at 1.064 µm. The ground-based station propagates a concentric and coaligned 488 beacon laser and a 1.06-µm relay laser through an 80-cm beam-director tracker. The beacon laser has a power of 2W and beamwidth of 57 µrad. The target scoring site (TSS), which is roughly 20 km away from the LSS, employed a scoring board composed of 37 sensors, enabling the determination of the accurate position of the received 1.06-µm signal. The TSS used 514 µm for its uplink, with the beacon laser having 4 watts of power and beamwidth of 76 µrad. In one phase of the experiment, the payload optics included a 150-mm-diameter hollow retroreflector developed and patented by Precision Lapping and Optical Inc. (PLOCI). The retroreflective mirror subsystem could allow the provision of a reference signal for the alignment and adjustment of the adaptive optical subsystem (AOS) of an optical ground station.

One such system design that could be used to attain a reduction of uplink distortion by means of an adaptive optics subsystem with a retroreflected mirror on board a satellite shown in Figure 8.8(b), uses a ground-based laser (GBL) at the optical ground station. As seen, it drives an uplink to the retroreflector and then comes down through the atmosphere as a reference for the AOS. The remediation of the uplink's atmospheric distortion enables, for example, the uplink command messages to be transported by means of the relay mirror to a spacecraft in deep space.

The alternative that needs to be weighed against the above system is shown in Figure 8.8(a). This system architecture consists of a reference laser on board the relay spacecraft, directing its output to the AOS of optical ground station and providing thereby the necessary compensation of the atmospheric turbulence.

8.5 High-Energy Deposition

Another planned experiment in which a mirror reflector structure could be useful is in the handling of a high-intensity laser beam [7]. For this application, the mirror would be made stronger and thermally more efficient. This is necessary as its function is to reflect a high-energy laser beam, referred to colloquially as HEL, from a ground-based source toward a target. Then it would deposit energy on the target in order to diminish it (Figure 8.9).

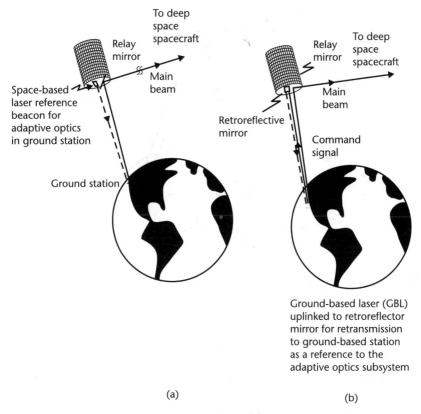

To deep
space
spacecraft

Relay
mirror

Main
beam

Space-based
laser reference
beacon for
adaptive optics
in ground station

Ground station

Relay
mirror

To deep
space
spacecraft

Main
beam

Retroreflective
mirror

Command
signal

Ground-based laser (GBL)
uplinked to retroreflector
mirror for retransmission
to ground-based station
as a reference to the
adaptive optics subsystem

(a)

(b)

Figure 8.8 (a) Use of relay mirror for command signal transmission to deep space (S/C); reference laser onboard satellite; (b) Use of passive reflectors on satellite with GBL uplinking reference signal toward retroreflector for downlink to the AOS. The compensated command signal is then transmitted to deep space (S/C).

Based on the success of the low-intensity laser used in the RME experiment, it is reasonable to assume that the HEL system could be used with an appropriately designed mirror to achieve a number of military missions, even in today's strategic and tactical environment.

A third application of the hardened articulating mirror in space is aiming a high-intensity laser, which is generated at a ground station or on an aircraft, at the mirror. The mirror reflects the laser pulses through the ocean toward a submerged platform. Returns from the target are collected bistatically by a receiver located on a separate aircraft or a low-altitude satellite. This system concept is shown diagrammatically in Figure 9.2. While this configuration is of course a ladar (laser radar), it may even be used as a low-data-rate communication system for an underwater vessel.

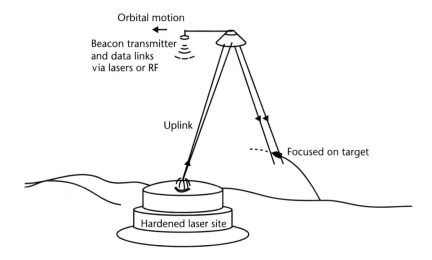

Figure 8.9 Ground-based high-energy laser aimed at a space-based mirror and directing the energy toward a target.

8.6 The Optical Westford

The Optical Westford, whose structural configuration is shown in Figure 8.10, is composed of a large number of very small mirrors (each roughly 10 times the wavelength of the impinging laser carrier).

The tiny spheroidal mirrors are released from the payload pouch of a spacecraft, at synchronous altitude, and are simultaneously observed between the transmitting and receiving telescope-equipped Earth stations. It is initially assumed that the large number of the small mirrors are bunched up as a spherical cloud. In fact, the system is like a "sphere containment of the mirrors" capable of reflecting a laser beam evenly in all directions, as with the case of a balloon. However, a beam from a primary source, whether it is an uplink from a ground station, an airborne platform, or a satellite astride the "cloud," when aiming at a particular spot on the sphere of little mirrors, will result in a reflected downlink to the ground station.

The reflective characteristics of the spherical bundle of little mirrors would initially be similar to the balloon, with the drifting of the mirrors at a differential rate. Those on the lower section would be moving faster than those on the upper section, making the drift swirl to a degree and with the spherical structure stretching and distending until the mirrors separate and move away from the initial laser spot. In a way, the drifting is quite similar to the RF version of the Westford described in Section 8.1.2.

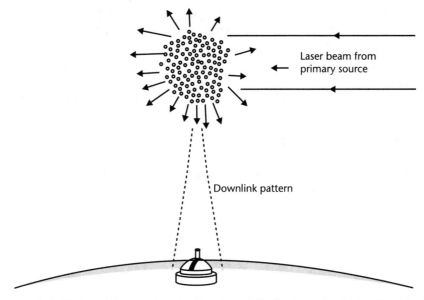

Figure 8.10 The Optical Westford: tiny spheroidal mirror, chaff-reflecting portion of a laser beam signal to a ground station.

It would appear that for short-duration military engagements occurring in spatial, ground-, and sea-based environments, the use of the mirror bunching in a spherical or any other bound shape could be useful in spoofing, feigning, and various other measures and countermeasures in the optical and infrared warfare genre. But the disadvantage of the mirror cloud is the drifting of the tiny mirrors, at different velocities, which leads to a short duration of the lifetime of the collective structure. This makes the useful engagement relatively short.

8.7 Additional Design Considerations

8.7.1 Use of GPS, Ground, and Retrodirective Mirrors to Locate Reflective Faces

The previous discussion is based on the assumption that the primary signal platform searches for the reflector structure (the secondary platform) with the aid of the GPS. Is also uses position data gathered from available ground stations and attitude data transmitted from the reflector satellite, in a direct or indirect way, to the primary satellite. This will enable it to point the laser beam center toward the center of the mirror face. Experimental development is leading to consideration of the use of retrodirective mirrors on the reflector satellite, enabling its attitude determination from the primary platform. The maximum level of power that may be used to ensure that no damage accrues on the secondary platform, as well as other details of such a signal configuration, may be drawn from [7]. This approach may be feasible, particularly if the return from the retromirror is modulated by means of an acousto-optic or electro-optic subsystem. The use of a mechanical jitter communication for the light modulation may also be useful.

8.7.2   Trades: Increasing the Signal Bandwidth

The techniques that can achieve a reference data rate of 10^9 bps are mode locking, modulating the pumping source or the switching technology. However, the current state of the art in terms of achievable data rate is hundreds of gigahertz and is limited, in terms of communications, by the detector design, not the source modulation. Techniques such as dense wave multiplexing (DWM) may enable each different color to be modulated by an information rate of, for example, 300 GHz. When 10 or more colors (frequency lines) are separately generated, they can be combined to produce 3,000 GHz into a single beam. This feasible frequency widening per beam could enlarge the reference data rate of 1.0 GHz used in this chapter by a factor of 3,000. Clearly, all the examples shown previously need to be appropriately scaled.

8.8   Summary and Concluding Remarks

Two passive reflective structures are described in this chapter. They are based on the RME program, which demonstrated in 1999 how a ground-based station transmitting an uplink beam to a reflector structure in space will reflect the signal toward another location on the Earth. In the experiment, the chosen wavelengths were two lines in the visible spectrum and one in the NWIR spectrum, and while the use of such an experiment provided emphasis for potential SDI needs, the interest to the communications community is in directing data flow from a signal source (or sources) to a receiver platform (or receiver platforms) on the ground, in the atmosphere, and in space.

Of the various reflectors that may be considered, the articulated space-based mirror and the Optical Westford System were selected for review in this chapter. Other configurations are also feasible but were found to be more costly and less reliable. In fact, the articulated mirror system was the most efficacious of all. The Westford, although less expensive than the articulated mirror, has a very limited longevity and can help only in certain types of tactical communications support.

There are several purposes in having passive reflectors: avoiding the need for another laser transceiver on board a satellite, reducing the cost of the intermediary space platform that transfers the signal to an Earth station, and, in unique applications, spoofing the enemy's communications.

The articulating mirror is also efficient and effective in terms of bandwidth, tracking, pointing, and focusing capability. Other versions of the articulating mirror, in particular those in which the mirror is hardened, will enable them to perform other functions. In particular, as is discussed in Chapter 9, the hardened mirrors will be able to focus high-signal energy toward a target from a ground-based high-energy laser (HEL) station. This reflective process includes a number of important applications: depositing energy pulses on selected targets, using ladar pulses (blue-green) to penetrate ocean water to search and identify submerged vessels, and also communicating, at a low data rate, to underwater vessels.

The reflective mirrors also have special utility, depending on the particular applications in which they are employed. In the main, however, it is clear that various kinds of balloons and portions thereof can mislead enemy communications in a number of strategic and tactical environments. With respect to the retroreflector mirror described in Chapter 9, there is the benefit of its use in the area of identification friend or foe (IFF) and in other communication sorties.

References

[1]   Dirks, J., S. Ross, A. Brodsky, P. Kervin, and R. Holm "Relay Mirror Experiment overview: A GBL Pointing and Tracking Demonstration," *SPIE,* Vol. 1482, 1991.

[2]   Sydney P. F., M. A. Dillow, J. E. Anspach, P. W. Kervin, and T. B. K. Lee, "Relay Mirror Experiment Scoring Analysis Scoring Analysis," *SPIE,* Vol. 1482, 1991.

[3]   Armstrong, J. W., C. Yeh, and K. E. Wilson, "An Earth-to-Deep-Space Optical Communications System with Adaptive Tilt and Scintillation Correction Using Near-Earth Relay Mirrors," JPL/Caltech, TMO Progress Report 42-135, 1998.

[4]   Aviv, D. G. "The MUGM (Miniaturized Unmanned Ground Based Mobile) Elements," *ARC Inc Reports,* 1992 and 1996, with briefings at Rand and Science Applications International Corp.

[5]   Reiger, S. H., "A Study of Passive Communication Satellites," The Rand Corporation, R-415-NASA, February 1963.

[6]   Martin, D. H., *Communication Satellites,* 4th Ed. Los Angeles: The Aerospace Press, 1986.

[7]   Henderson, W. D. "Integrated High Energy Laser for ABM Applications," *AIAA,* July 1988.

[8]   Wilson, K. E. (of Caltech), and P. R. Leatherman, R. Cleis, J. Spinhirne, and R. Q. Fugate (of Kirkland Air Force Base, Albuquerque, NM), "Results of the Compensated

Earth-Moon-Earth Retro-Reflector Laser Link (CEMERLL) Experiment," JPL-TDA Progress Report 42-131, November 15, 1997.

[9]   Creamer, N. G. "Multiple Quantum Well Retro-Demodulators for Spacecraft-to-Spacecraft Laser Interrogation, Communication and Navigation," 15th Annual AIAA/USU Conference on Small Satellites, 2001.

Appendix 8A

To indicate the robustness of the mirror in a space type of communications subsystem, another experiment, based on NASA spatial development, should be considered [8].

Through NASA's *Apollo 15,* a retroreflective mirror array was placed on the Moon's surface at Hadley Rille, with technical personnel located at Kirtland AFB coordinating the transmission and collection of the data and performing the data analysis. The transmitter was located at Kirtland's Starfire Optical Range (SOR). It had a 1.5-m aperture and the receiver, also at SOR, had a collector aperture of 3.5m. JPL provided the polynomials to locate the retroreflectors on the Moon, which were then used to generate the tracking algorithms for both transmit and receive telescope antennas.

With atmospheric correction by the AOS at the receiver aperture, the number of received photons per pulse was more than 100. The laser was an Nd:YAG generating 1.06-μm signal, and the reference laser for the AOS was a laser guide star. The experiments were performed from March through September 1994, through the first and last quarters of the Moon.

As far as Chapter 8 is concerned, this Earth–Moon experiment shows yet another example of the utility of a passive mirror system as an important adjunct of laser communications and other laser systems applications, such as energy deposition, ladar, and countermeasures.

9

Unique Applications
of Laser Communications

9.1 Introduction

This chapter describes a number of special applications of laser communications covering combined RF/laser communications, underwater communication, ladar (laser radar), and laser communication subsystems for microsatellites and nanosatellites. However, the planned laser link from the Moon to Earth, as well as a laser communications system from Mars to Earth, which have been studied in detail by NASA, have been canceled because of cost issues.

We first discuss the ability to achieve communications in combined RF/laser bands using the same telescopic antenna, with emphasis on the robustness of command and control applied in various tactical and strategic communications. In this manner, where convenient in terms of the inclement environmental conditions, RF may be used, and in fair weather conditions, laser bands are used. The selection of the bands may take place during the communication period between the two platforms.

In terms of the unique applications outlined in this chapter, examples are given for laser communications between a satellite and a submerged submarine, in depths down to 100m. Also, by virtue of potentially attainable high-energy laser equipment, a unique laser radar system is conceptualized. Called a space-based detectability and identification of submersibles (SBDIS) system, it will be capable of delivering, from a ground-based station, high-energy laser pulses aimed at a space-based mirror, scanning regions of interest of the ocean. The reflected emanations from a submerged vessel will be picked up by an aircraft or an LEO satellite, and upon processing, the signal may identify an underwater target of interest.

The next potential of underwater laser communication is the satellite-to-submarine downlink and the submarine-to-satellite laser communication. The various loss components of the dowlink/uplink using blue-green line are presented.

Apart from the SBDIS and submarine laser communication to satellite (SLCSAT) applications, a return link system (RLS) is also considered. In the RLS application, a submarine, while receiving command messages by ELF/VLF, may need to respond by transmitting messages back to headquarters, thereby indicating that the command message has been received and authenticated. As will be shown, the return link signal may be propagated in the visible or near-infrared laser by means of a small, stabilized, trawling platform on which a laser transmitter is bolted.

The next section of this chapter discusses the deployment of a steered agile laser transceiver (SALT) on a nanosatellite whose weight is assumed to be less than 20 lb. (~9 kg). The SALT, as it is now being developed in the laboratory, is planned to be a 1-inch cube weighing 1 to 2 grams and might within a few years have the capability of transmitting more than 1.0 Mbps for a distance of 1,000 km, at a better than 20 dB S/N. Unique developments of nanotechnology are being included in the SALT cube design. The emphasis in the advanced SALT version will be to extend performance to a range of 10,000 km with up to 100-MBps bandwidth. However the weight of the advanced SALT (ASALT) subsystem is likely to increase by a factor of 5.

Another communication system concept known as the retroreflective communication subsystem (RRCS) is also described in this chapter. In this design, a continuous wave (CW) laser onboard a "mothership" satellite is used to communicate with a nanosatellite. Onboard the nanosatellite is a solid retroreflector mirror system, which, upon receiving the interrogation beam from the mothership, will return the signal back to it. But before being reflected, the beam will go through a multiple quantum well modulator (MQWM), which is driven by a signal source on board the nanosatellite. In this manner, the signal received by the mothership will contain the intelligence generated by the signal source, such as a camera, for example.

This chapter also provides overview calculations of the signal power budget of the laser communication system between the Moon and the Earth, between Mars and Earth, and also between the Mars Reconnaissance Orbiter (MRO) and Earth. While these links have been studied in detail, the lack of funding has regretfully led to cancelation of the three programs. Microwaves will be used to transmit data from MRO and from the Moon to the Earth.

This chapter also mentions the planned transformational communication architecture (TCA), in which wideband laser links are considered between intersatellite links. These satellite network–centric nodes are also combined with RF and laser links for uplink and downlink from ground-based, airborne, and sea-based platforms. In the 5-GENIN System discussed in Chapter 7, the synchronous backbone may become the fundamental branch of the TCA.

9.2 Combined RF and Laser Telescope Antenna

Because the communications system must perform in different weather conditions, a telescope antenna was conceived for the purpose of being able to operate in cloudy and rainy weather as well as in clear weather. Such an antenna would be further advantageous when deployed on physical platforms that are space limited. That is, having one antenna structure instead of several and using the multiband antenna for radar as well as communications has many benefits.

The basic multiple-sensor coaxially configured antenna system is shown in Figure 9.1. As indicated, two different RF bands are used on either side of the Cassegrain antenna. RF/f-1 is behind the parabolic section, and

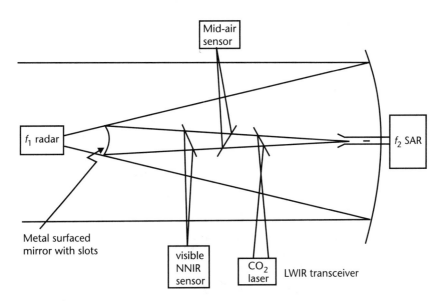

Figure 9.1 Multiple communications links employing laser (visible, NWIR, MWIR, and LWIR) and RF (mmW and microwave) bands using a coaxially configured Cassegrain antenna system.

RF/f-2 is behind the hyperbolic section. Both bands may operate simultaneously, as communications links or as radar systems.

The laser transceivers in the visible band and NWIR are extracted along the central axis of the antenna, followed by a midwave infrared (MWIR) sensor link and a long-wave infrared (LWIR) communication transceiver. All the four laser bands may be utilized either as carriers in a communication system or as a ladar system.

The multiband RF/laser antenna system is particularly useful onboard a ship where space is at a premium. In other applications, the available lines will enable a variety of visible and infrared imaging sensors (together with communications support), to collect a large amount of observational data, with high antenna directivity.

9.3   Space-Based ASW to Achieve Detectability and Identification of Submersibles

The SBDIS system (Figure 9.2) is based on the integration of a ground-based, space-based and airborne platforms operating in system unison, in the following manner.

The high-energy pulse laser, which is blue-green, is generated at a ground-based station. Its optical antenna is aimed at a space-based mirror, of the type described in Chapter 8. This mirror scans a particular ocean area of

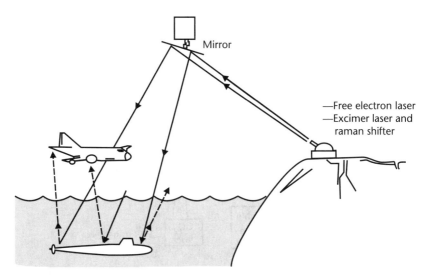

Figure 9.2   Space-based detectability and identification of submersibles system using bistatic configuration.

interest, searching for returns from an underwater vessel. The returns in the visible spectrum (blue-green) are picked up by a sensor on a LEO satellite or by an airplane or a high altitude airship that hovers about a region of the ocean being scanned.

The high-energy ladar system used to detect and identify targets at their typical operational depth will provide, upon processing of the return signal from their hulls and its extensions, unique underwater target returns, with a particular ladar cross-section (LCS). In today's technology there would appear to be two possible locations for the high-energy blue-green laser and optics. The first is on an airborne platform, and the second, which is considered most appropriate from a reliability and operational aspect, is deploying the HEL system in a ground-based station and having its beam aimed towards a space-based mirror [1], for reflection through the ocean.

9.3.1 Blue-Green Laser System Design Features

In this design we pump the Nd:YAG at high efficiency with laser diodes. They are grown with metal organic chemical vapor deposition (MOCVD) on a GaAs substrate. The output of these diodes is at 0.807μ, which is an Nd:YAG absorption line. A parallel program to attain a broader absorption band is the Gallattium Gallium Garnet (G_3), and GSGGH ($Gd_2Sc_2Ga_3O_{12}$), which, when codoped with Nd^{3+} and CR^{3+}, will increase the pumping efficiency. The pumped Nd:YAG output going through a nonlinear converter will produce a 0.48-micron blue-green line.

The wavelength control can be achieved with diode selection and temperature control. As will be seen, the attenuation through seawater is very sensitive to wavelength, hence controlling the wavelength is crucial when trying to reach the submerged targets.

9.3.2 Experimental Approach to Achieve the SBDIS Lidar

To determine the feasibility of remotely sensing ocean internal waves with airborne pulsed blue-green laser, a ladar internal wave detection experiment (LIDEX), may be considered.

The results of such an experiment could be useful because of the coupling of the submarine's energy into thermoclynes that separate warm surface water from colder deep water. The thermoclynes oscillate, and can be detected by blue-green lasers that can penetrate several hundred feet down through the seawater. The energy coupling produces a "gravity wave" in the thermoclyne that can be detected by laser radar onboard an aircraft or an LEO satellite.

The basic equation for the received laser pulse as a function of the transmitted pulse, sourced at a LEO satellite or an aircraft platform (Figure 9.3), transmitted toward the ocean containing the submarine target, and then reflecting from the target through the water and onward to the receiver aperture, is given in (9.1).

As seen, the received energy per pulse is equal to the transmit pulse divided by the steredian angle, which is the beamwidth squared of the transmitted angle, times the losses in the atmosphere, cloud-to-water transmittance, air-to-water transmittance, transmittance through water, laser cross section of target and angular scattering, return transmittance through the water, water–air interface, transmittance through air and through clouds, and also the angular scattering from the ocean surface and on to the area of receiver aperture. Expressed in terms of unique components, the received energy per pulse is

$$
\text{E received/pulse} = \text{E transmitted/pulse} \div \pi/4 \left(\theta_t\right)^2 \left[\tau_{atm}\tau_{cloud}\tau_{cw}\right]
$$
$$
\left[\tau_{aw}\tau_w\right] \cdot \left[\sigma \cdot \text{angular scattering}\right]\left[\tau_w\right]\left[\tau_{wa}\right]\left[\tau_{wc}\right]\left[\tau_{cloud}\tau_{atm}\right] \quad (9.1)
$$
$$
\left[1/\pi\left(\theta_t' R^2\right)\right]\left[A_r\right]
$$

where

τ_{atm} = atmospheric transmittance
τ_{cloud} = cloud transmittance

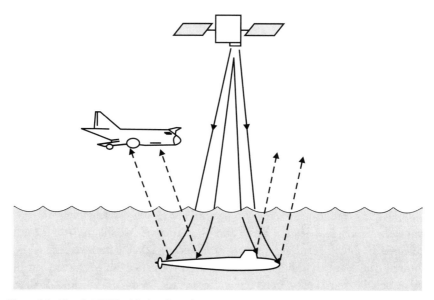

Figure 9.3 Bistatic LADAR with aircraft receiver.

τ_{cw} = cloud-to-water transmittance
τ_{aw} = air-to-water interface transmittance
τ_{w} = water transmittance
θ_{t} = beamwidth of transmitted beam
θ'_{t} = angular scattering bound from the ocean surface
σ = LCS of target
A_{r} = area of receiver aperture
R = distance between laser and entrance into ocean

This equation may be used when transmitting a laser pulse from an LEO satellite or an airborne platform.

9.4 Submarine Laser Communication to Satellite Concept

In examining the SLCSAT concept, we start from the submarine laser transmitter and enter the loss field due to the absorption and scattering by the seawater followed by beam spread due to changes in the index of refraction and to suspended biological particles and fading. Additional losses are due to reflection losses from the water-air interface, sea state loss, and the atmospheric losses and space losses. The loss layout is shown in Figure 9.4.

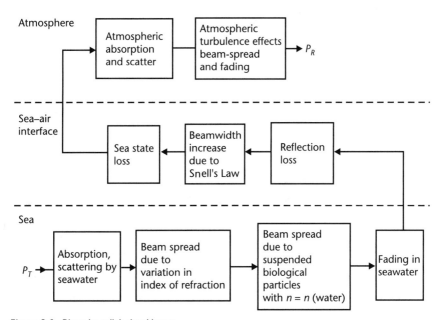

Figure 9.4 Direct laser link signal losses.

A key indication of the best wavelength line to use in the transmission of signal through sea water, in terms of its extinction coefficient has been measured in different parts of the world. As shown in Figure 9.5, the lowest extinction coefficient is in the 450 millimicron to 475 millimicron subregion.

Shown, however, are six different sea locations in the world's seas from the Galapagos to the South Baltic Sea in which the extinction coefficients were measured [2]. The attenuation based on an extinction coefficient of 0.12 per meter at 525 millimicrons is given in Figure 9.6 at 100-, 200-, and 300-foot depths (yielding ~50-dB loss).

Additionally there are the loss components because of the small amount of seawater scattering $\leq 10^{-3}$, due to refractive effects from globules having a large index of refraction variation and also due to suspended biological particles with an index of refraction close to that of seawater ($n = 1.43$).

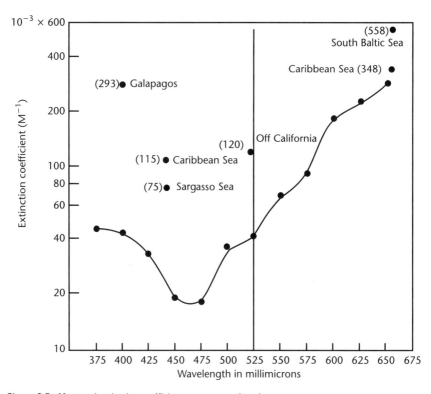

Figure 9.5 Measured extinction coefficients versus wavelength.

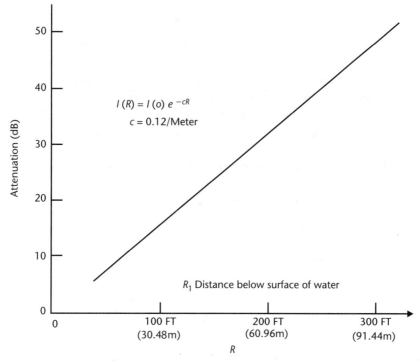

Figure 9.6 Attenuation through seawater at c = 0.12/m.

9.4.1 Loss Due to Beam Spread Resulting from Seawater Turbulence

The signal loss due to beam spread from seawater turbulence [3, 4] may be expressed in the following way:

$$\text{normalized beam spot} = (P_t / P_o)^2 \tag{9.2}$$

where

$$P_t^2 = \omega_o^2 \left(1 + R^2 / k^2 \omega_o^4\right) + \varepsilon R / 3 \tag{9.3}$$

indicates "beam cross-section area" (radius squared) in a turbulent medium, and

$$P_o^2 = \omega_o^2 \left(1 + R / k^2 \omega_o^4\right) \tag{9.4}$$

Equation (9.4) indicates "beam cross-section area" in the absence of turbulence. Also, ε, the measure of index of refraction variance relative to the radius of globules may be expressed as

$$\varepsilon = \left\langle \Delta n^2 \right\rangle / a_r \qquad (9.5)$$

where

a_r = radius of globules (orders of 10s of centimeters)
$\langle \Delta n^2 \rangle$ = average value of variance of index of refraction
ω_o = radius of Gaussian cross-section of laser beam
k = $2\pi/\lambda$

The power loss due to beam spread from seawater turbulence is then

$$\text{Power Loss} = -10\log\left(P_T / P_o\right)^2 = \qquad (9.6)$$
$$-10\log\left[1 + R^3\varepsilon / \left\{3\omega_o^2\left(1 + 2/k^2\omega_o^4\right)\right\}\right]$$

A plot of (9.6) is shown in Figure 9.7 for different normalized spot sizes as a function of distance in the water, from the submarine's transmitter, with beam cross-sections and ε as parameters. As seen, at a 100-m depth from the sea surface, the normalized spot size due to turbulence would be 11.

9.4.2   Beam Spread Due to Suspended Biological Particles

An additional signal loss of the laser beam when propagating from the sub-based laser transmitter, through seawater, due to the suspended biological particles, was outlined in Section 9.4. For example, the beam cross-section at distance R may be expressed, starting from (9.7), as

$$\text{power loss} = -10\log\left(P_p / P_o\right)^2 \qquad (9.7)$$

where

P_p = signal power through suspended bioparticles
P_o = signal in the absence of bioparticles

Then, the beam cross-section due to bioparticles at a distance R is expressed as follows:

$$\pi P_p^2 = \pi\left\{\omega_o^2\left(1 + R^2/k^2\omega_o^4\right) + 8/3R/R_c\left(R/kc_p\right)^2\right\} \qquad (9.8)$$

where

R_c = small-angle single scattering length

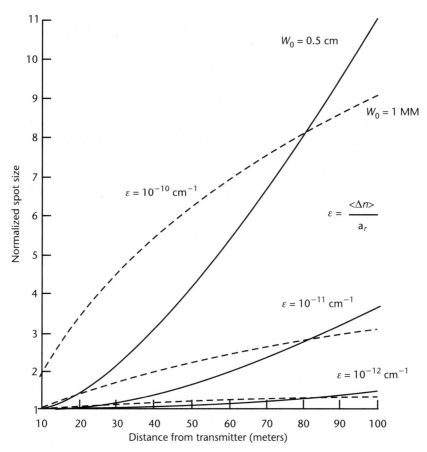

Figure 9.7 Normalized spot size in turbulent seawater.

$$R_c = 2n_o^2 / \left[\left(\pi^{1/2} \right) f \mu^2 k^2 a_p \right] \qquad (9.9)$$

n_o = mean index of refraction of the water medium
f = relative volume concentration of the suspended particles
μ = difference between the refractive index of the scattering particles and that of the water in the absence of the suspended particles

Equation (9.9) is valid for values of a_p large compared to the optical wavelength. This indicates $a_p \geq 10^{-4}$ cm.

A summary matrix of signal loss due to extinction and beam spread due to turbulence and suspended biological particles is shown in Figure 9.8. As seen, at 100 ft below the surface of the sea, the extinction loss is 16 dB, and it goes up to 48 dB at 300 ft. The beam spread due to turbulence is 2 mm (5-dB loss) at 100 ft and 18.6 dB at 300 ft.

Depth	100 FT (30.48M)		200 FT (60.96M)		300 FT (91.44M)	
	Radius of spread	Power loss	Radius of spread	Power loss	Radius of spread	Power loss
Extinction loss		16 dB		32 dB		48 dB
Beam spread due to temperature variance $\omega_0 = 1$ mm $\varepsilon = 10^{-10}$cm^{-1}	2 mm	5 dB	5.6 mm	15 dB	8.7 mm	18.6 dB
Beam spread due to suspended particles $\omega_0 = 1$ mm $\varepsilon_p = 10^{-4}$ cm $R_C = 10$ m	250 mm	33 dB	300 mm	36 dB	410 mm	37 dB
$\omega_0 = 2$ mm $\varepsilon_p = 10^{-4}$ cm $R_C = 10$ m	430 mm	37 dB	600 mm	39 dB	800 mm	40 dB

Figure 9.8 Summary of signal loss due to extinction and beam spread.

Due to suspended biological parameters, the beam spread, with an initial $\omega_0 = 2$ mm, is 43 cm (37-dB loss) at 100-ft depth and 40-dB loss at 300-ft depth. The total loss at 100-ft depth is ~60 dB, and 107 dB at 300-ft depth.

In terms of the interface loss, it is typically another 3-dB loss plus the sea state loss, and of course we have another ~300 dB as the space loss. The atmospheric fading due to turbulence is taken for this example to be an arbitrary 7 dB. (It should be recalled that since the beam has already been significantly spread in sea water and at the sea/air interface, there would be a relatively small broadening due to atmospheric turbulence). But as noted in Chapter 6, the loss due to clouds could be much higher than 7 dB, so that a range of cloud losses should be considered when estimating the signal loss.

The LEO satellite receiving the laser signal would need to relay the signal to a command center at CONUS, or the LEO satellite would relay the signal to an airborne command post. Several LEO satellites receiving the laser signal from the submarine will be scanning particular ocean areas and tracking the signal. A typical aperture size of an optical antenna of a LEO is 60 cm with a 22° FOV. The Cassegrain antenna will include an adaptive optical subsystem, and its hyperbolic secondary lens will be the scanning element.

9.5 Approximation of Laser Signal Loss in Satellite-to-Submarine Communication

The essence of loss components in a laser signal going from a satellite to a submerged submarine is shown in Figure 9.9. As seen, the space loss is typically ~300 dB, the loss term when going through another "model" cloud is assumed to be 4–14 dB, followed by the air/seawater interface loss. Added to these would be the losses in the water, which include the product of the diffusion attenuation and depth, yielding the loss in the sea water to be 5–50 dB, to a depth of 300m.

9.6 Return Link Communication

Upon receipt of a command message by the submarine, especially the Polaris missile sub, it is very important to transmit a message back to the command center, notifying it that a message of a particular type has been received and identifying within a coded field the name of submarine and other data. To

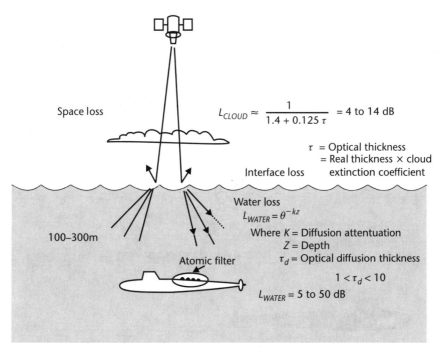

Space loss

$$L_{CLOUD} \approx \frac{1}{1.4 + 0.125\,\tau} = 4 \text{ to } 14 \text{ dB}$$

τ = Optical thickness
= Real thickness × cloud extinction coefficient

Interface loss

Water loss
$$L_{WATER} = \theta^{-kz}$$
Where K = Diffusion attentuation
Z = Depth
τ_d = Optical diffusion thickness

$$1 < \tau_d < 10$$

100–300m

Atomic filter

$$L_{WATER} = 5 \text{ to } 50 \text{ dB}$$

Figure 9.9 Approximation of loss relationships for satellite-to-submarine communication.

avoid transmitting an RF message, which would require an antenna to be raised or floated, a brief laser signal is transmitted from a very small "vessel." Such a boat would be of a size of less than ~46 cm in width and ~30 cm in diameter; it would be covered with a radome bubble of the type shown in Figure 9.10.

As indicated, a cable connecting the submarine's sail to the floating bubble is unwound, and the laser signal is transmitted via this cable to the telescope. The latter is bolted to the small stabilized platform on the small vessel. This signal, which is short in duration, is propagated to an LEO satellite or even a synchronous satellite. Dummy floating bubbles may be used for passive countermeasure purposes.

9.7 Interplanetary Laser Communication

In the next two sections, we review the essence of the desired laser communications between Moon and Earth and between Mars and Earth. Because of the atmospheric losses, it might be necessary to have the downlinks toward the Earth be terminated at one of the three synchronous satellites orbiting the Earth, which are called the synchronous backbone in Chapter 7.

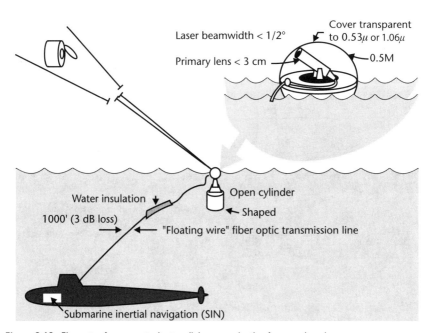

Figure 9.10 Elements of a conceptual return-link communication from a submarine.

Should the synchronous backbone be available when the Mars and Moon laser links are in place, this vital option could play an important role in providing continuous communication in a time-sensitive environment.

From the synchronous backbone, the signal will be downlinked to an OGS that is situated in a clear weather environment. The latter OGS, as with the other ground stations, will be connected by means of a worldwide fiber network to the central command station. Based on current calculations, it is anticipated that the total bandwidth of the laser downlink will be of the order of 1.2 Gb from the Moon to the Earth station and 10M/sec, at ~1 AU, for the Mars downlink to an Earth optical station.

At this time it should be pointed out that the optical link design, from an optical transmitter located on the surface of Mars that is aimed at the surface of the Earth, is but conceptual. In fact, the planned robotic laser beam operations from Mars to Earth, as well as from Moon to Earth, have been canceled. But it is desirable to determine what the various components of the SPB are and anticipate that eventually such laser communication systems will be deployed. Based on this premise, a number of communications components and subsystems have been selected by NASA and a detailed evaluation of S/N has been made. Clearly, the basic difference between laser communications between satellites in space and Mars-to-Earth planetary links is the vast distance between Mars and Earth throughout their orbital cycles, typically in the range of 1–2.5 AU. Moreover, there are the accumulated background noise photons, which increase with distance, relative to a constant number of signal photons.

However, in an example of a link from Mars with the aperture design of 20 or even 30 cm, it would be feasible to close the link, given an optical transmitter with more than 3w average laser power. And as there are few limitations on the Earth optical receiver, a 5m or even a 10m aperture should provide a useful signal margin to close the link. The support of the Mars-to-Earth links to achieve 1–10 MBps is discussed in the next section.

9.7.1   Feasibility of Laser Communications Between Mars and Earth

NASA/Goddard with support from JPL and MIT-Lincoln Laboratory have been leading the studies of Mars-to-Earth laser communication systems. While the program to implement the laser link system has been canceled, the key design features of the transceiver have been detailed and performance feasibility has been determined. In fact, calculations have shown that a 20-cm aperture telescope on Mars's surface and about 3W average laser power, and a 5-m aperture telescope on the Earth's surface, will support a bandwidth of 10 MBps, at 1 AU (~93 million miles). This link should be of particular interest when the distance between Mars and Earth is increased to more than 2 AU. Using the SPB developed in Chapter 2, it is clear that the bandwidth will be reduced to ~1 MBps. Apart from the distance increase, there are the added noise photons that contribute to the reduction in S/N.

Clearly, improvements in bandwidth performance could be had with a larger telescope on the surface of Mars. For example, one might consider a 1-m or larger aperture telescope. But the cost and complexity of transporting large size optics from Earth to Mars and deploying and adjusting it on Mars would be prohibitive. This is because the current state-of-the-art lift and transport technology and robotic reliability is not sufficiently developed. However, current improvements in laser power output technology may make it easier to achieve loop closing. This potential is now being evaluated.

As a backup, an X-band transmitter on the surface of Mars may be considered. For example, with a 28-cm dish and 15W of average RF power, plus a 70-m antenna dish on the surface of the Earth, an equivalent signal bandwidth would be achieved. The stations on the Earth could be located in the United States (California), Australia, and Spain.

A second possible laser design link, that from the MRO to Earth, would be comprised of a 30-cm aperture telescope on the MRO and a 3-W average power laser. The telescope aperture at the Earth's ground station would have to be 10m, and the calculated bandwidth would be 65 MBps at 1 AU. The MRO is planned to be orbiting Mars with a circular orbit of about 600 km–800 km.

However, since the actual transmitter is a microwave communications sytem at Ka-band instead of a laser beam between MRO and the Earth, we are likely to have, as a key component, a 1-m dish at Ka. As an alternative to the Ka transmitter, an X-band transmitter with a 3-m dish antenna on the MRO is now being used.

9.7.2   Further on the Mars-to-Earth Laser Communication Potential

In any of the Mars-to-Earth communications, it should be recognized that while the laser communications has advantages over the microwave communications modality, the lower cost, higher reliability, and greater experience of microwaves systems over the laser communication systems gives significant advantages to the RF. While the laser links have not been on a "go" basis, components have been fabricated and calculations are being repeated for different conditions with varied simulations.

In preparing this book, it was recognized that the information from NASA agencies regarding laser versus RF communication efforts for the Mars-to-Earth links was fluid, because of the funding, roboticity, and reliability issues. However, what is paramount is that the physics and engineering are sound and that the multiple calculations shows the feasibility of the optical links from Mars to Earth and also from the more distant planets to Earth.

9.8 Proposed Laser Communication Between the Moon and Earth

The planned laser communications from various sensor suites on the Moon could achieve a high resolution video as well as a detailed instruments read-out to the Earth. The advantages of using laser beam communication are as before; the transmitting and receiving terminals are much smaller and lighter in weight then their RF equivalents, and with higher security.

As described by Paul Blasé [6] of the Artemis Society International, an example of a Lunar laser transceiver was conceived several years ago by Astro Terra Corporation of San Diego, CA. It consisted of a physical package of less than 10 in^3 (~16.4 · 10^3 cm^3) weighing 31.6 lbs (~14.4 kg) with a telescope and gimbal, electronic module, and deployment mechanism. In one design, it would use a 13.5-cm diameter Schmidt-Cassegrain telescope as a receiver (satellite and ground station having the same receiver design). In another proposed design, the SBPB is given in Table 9.1.

The system would consist of eight solid lasers, each 125 milliwatts, at 810 nm with 500/1500-μrad divergence. Four of the communication lasers would form one 600-Mb/sec channel, transmitting with right-hand circular polarization, and the other four would form the other channel, transmitting with left-hand circular polarization. Thus, the total of 1.2 Gb/sec would be transmitted [6].

Regretfully, the Lunar-to-Earth laser communication system was also canceled. The Astro Terra Corporation, which performed the preliminary work on the laser communication system, regretfully had their work terminated. However, Astro's research and development projects should be useful in future planetary communications. In the meantime as with the Mars communication system, the Lunar communication system will also be in the microwave band.

9.9 The Microsatellite or Nanosatellite

Because of the lower cost and reliability consideration, the typical weight of a satellite has been reduced over the past 20 years, going, in many applications, from well over 1,000 lb (~455 kg) down to less than 200 lb (~90 kg), and even down to 20 lbs (~9 kg). The developing technology of composite material was the first step in lowering this weight level, followed in the past 10

Table 9.1

Proposed Design Parameters of a Laser Link from Moon to Earth

Transmitter	
Laser power	1.0W
Beam divergence	1.0 mrad
Telescope magnification	300
Atmospheric transmission (estimated)	0.90
Beam divergence after telescope	3.33 μrad
Receiver	
Range (Moon-to-Earth station)	384,790 km
Spot diameter	1,282.6m
Power density	$6.97 \cdot 10^{-7} W/m^2$
Telescope window diameter	0.5m
Window area	$0.196m^2$
Optical efficiency	0.60
Receiver power	$8.21 \cdot 10^{-8} W$
Required power	$4.8 \cdot 10^{-8} W$
Margin	2.3W

years by improvements in a variety of nanotechnology subsystems and components [7].

In this section, we examine several achievements attained with microsatellites and then look at the potential inclusion of the advanced design of a small-size laser transceiver to enable intersatellite laser links between any two small space platforms.

A 6.5-kg nanosatellite, known as the *SNAP-1* and developed at the University of Surrey, was launched successfully in June 2000. It supports a microsized GPS receiver system, a camera and associated optics, a computer, and propulsion and attitude control subsystems. The primary payload consists of a machine vision system (MVS), which enables the inspection of particular spacecraft in orbit. The MVS is made up of three ultraminiature wide-angle complementary metal oxide semicontuctor (CMOS) video cameras and one narrow angle CMOS video camera with an onboard processing computer.

The MVS can also be used to shoot, with medium resolution (~500m), targets on the ground from an altitude of 650 km at a near-polar orbit. For example, *SNAP-1* was able to shoot pictures of a Russian military satellite in orbit and later on, rendezvous with a companion satellite named the *Tsinghua-1* microsatellite, after various maneuvers. It used its onboard GPS receivers and a tiny butane propulsion system. Future nanosatellites will be used for space inspection duties, examining the International Space Station, supporting small space science instrumentation, and with intersatellite laser communication, enabling formation flying and attaining measurements that require spatial and interferometric diversity. As we have said, the need for laser communication rather than microwave for intersatellite links is that the beam narrowness and the wideband capacity are more easily attained. Moreover, with laser communications it would be more difficult for anyone who is planning to interfere with the communications link to get into any sidelobes of the laser beam with jamming signals. (See Figure 1.2.)

9.10 The SALT System

The SALT system is another system now being developed. Conceived by Professor Kristofer Pister [7] and his team at the University of California in Berkeley, it may be selected as the communication subsystem for the nanosatellites.

The main features of the SALT are its bidirectional steered communication subsystem, which might fit into a cubic centimeter volume. It will consist of gyrostabilized minilaser turrets and a 1-mbps to 1-GBps CMOS imaging receiver. Its weight might be only 1 gram. Its power consumption is currently assumed to be 0.1–5W, depending on the data rates. The imaging receiver will be able to receive dozens of 1.0-MBps to 1.0-GBps laser signals simultaneously.

The anticipated SALT system application to perform effectively as a laser transceiver onboard a satellite for ILL is requiring additional research and development at the University of California. As an example, a composition of semiconductor laser beams by means of Bragg beam benders [8], which have been developed by Stoll and Garmire at the Aerospace Corporation, might be used in the advanced SALT configuration.

Even before the ILL function, there are potentially other applications, which could exploit the agility, small weight, and low power requirements of the SALT. These would include platforms such as microair vehicles (MAV) and ground-based force elements, such a swarm of MUGMs.

A summary of the projected features of the gyro-stabilized steered agile transceiver (Figure 9.11) is given below.

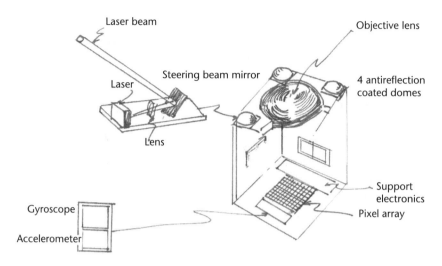

Figure 9.11 Steered agile laser transceiver unit, simultaneously communicating at 1–100 MBps with multiple transceivers.

- Size: ~1(cm)^3
- Weight: 1g to 10g, depending on data rate
- Power: 0.1–5W, based on data rate
- Range: 10 km in daylight for 1 MBps and 0.6 km for 100 MBps
- Data transmission rate: up to 1 GBps from each of 4 laser turrets
- Data reception rate: 1 MBps–100 MBps, simultaneously, from each of up to one-hundred transmitters
- Acquisition time: 0.9 ms average at 1 km, less than 50 ms at 10 km
- Tracking: roll rates of 1,000 rad/sec (166 rps)
- Desired goal: Inexpensive and mass produceable commercial, off-the-shelf (COTS) CMOS and commercially available MEMS

Following successful results of the basic SALT development, the Advanced SALT when expanded beyond the size of 1(cm)^3, should, within five years, achieve a higher optical output power and a larger mirror diameter to provide the capability of a longer range, at a low BER for OOK modulation. This should enable communication between two large SALT transceivers deployed on separate satellite platforms. The components in the SALT system to be enhanced would be the semiconductor laser with its optics and steering mirror; the gyroscope and accelerometer; pixel array; application-specific integrated circuit (ASIC) controller; objective lens; and antireflective dome coating.

As more experience is gained with the SALT subsystems, they would also be used between two ground stations. This would be followed by the deployment of SALT between aircraft and ground-based stations. Such links will be expanded by the ASALT system design, enhancing its performance to achieve intersatellite communication with distances of up to several 1,000 km. As emphasized in Chapter 2, this would require higher laser power and larger optical apertures. Both are now being developed by several organizations. These are principally Ball Brothers and also the federally contracted research laboratories: MIT Lincoln Laboratory and JPL.

9.11 The Retroreflective Communication System

The concept described in this section is based on a mothership satellite communicating via lasers with a microsatellite or a nanosatellite platform. (The title "mothership" is given to the satellite that is transmitting an interrogation signal, a basic command signal, or a relay communication signal to selected satellites. In our example, these are very small satellites.) The small platform collecting the interrogating laser signal responds via a retroreflective mirror, on which modulated information is superimposed. The data generated by the microsatellite is modulated on the return signal by means of a transmissive multiple quantum well (MQW) modulator [10]. More details of the system concept onboard the interrogated microsatellite are shown in Figure 9.12.

As seen, an interrogation CW signal arrives from the mothership at the microsatellite. This signal gets modulated by an information source and then reflected through the solid retroreflector mirror, back to the mothership satellite.

When the retroreflective communication system (RRCS) is fully developed, it will be deployed and tested on a satellite that has been identified in this discussion as the mothership and also on a selected microsatellite. The communication performance between the two platforms will be measured. The data rate may be of the order of Gb/sec. Depending on the coding of the interrogating signal, the system may also function as an identification friend or foe (IFF) system. This may open up a new field of application for the RRCS.

The special advantage of the RRCS is that we do not need another laser onboard the microsatellite (or nanosatellite) to achieve "back and forth" communications. The design takes full advantage of the retroreflective mirror, which is a totally passive device. Of further advantage of the RRCS is its hardness to proton radiation. For example, experiments have shown that even under bombardment of 20-MeV protons, up to a total exposure level of $6.4 \cdot 10^{10}$ protons per cm^{-2}, no degradation in the

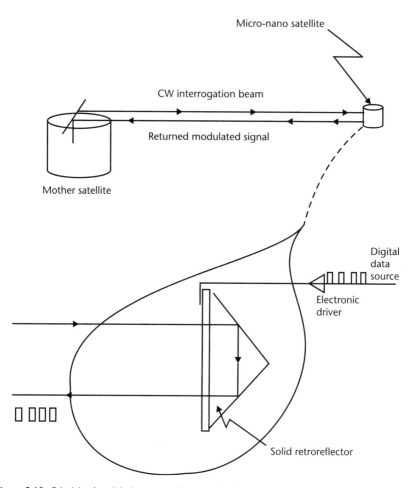

Figure 9.12 Principle of modulating a retroreflector mirror by means of a multiple quantum well modulator (CW may be replaced by a pulse-coded signal).

InGaAs/AlGaAs modulator performance was detected. The simulation of
this environment was shown to bear no effect on the key components of
the RRCS [11].

While it may be considered feasible, the RRCS and certainly the SALT
are not yet ready for spatial deployment and particularly not yet ready for
platforms that are separated by several thousand kilometers. However, it is
estimated that within a five-year period of continued research effort, these
systems could be deployed on microsatellites and nanosatellites and also the
UAVs and, eventually, the MUGMs.

Let us now examine some of the data link aspects of the RRCS. First,
the MQW technology should allow data rates of tens of megabits per second.

With diffraction-limited optical power, the retroreflected signal from the microsatellite or the nanosatellite to the mothership would scale as

$$P_{rec} = P_{laser} D_{retro}^4 D_{rec}^2 / \theta_{div}^2 R^4 \qquad (9.10)$$

where

P_{rec} = received power at the mothership
P_{laser} = laser power aimed from the mothership to the small platform
D_{retro} = diameter of the modulator retroreflector on the small platform
D_{rec} = diameter of the receive telescope on the Mothership
θ_{div} = divergence of transmitter laser beam
R = distance between the mothership and the microsatellite

As expected, and as shown in (9.10), the key dependence of the link is on the range between the platforms and the diameter of the retroreflector.

9.12 Summary and Concluding Remarks

Chapter 9 starts with a description of an antenna telescope structure that accommodates both multiple RF and multiple laser lines and may be used for communications and for radar. It thus provides service in clear and also in inclement weather.

The chapter continues with emphasis on potential applications of laser communications to a satellite from a submarine and vice versa. We also describe the use of high-energy laser pulses that could be used as a ladar, to detect submerged vessels. Although this book is about laser communications, ladar application is included because the resulting signal response requires communication to a command center. Also, under special circumstances when the targets are underwater, communicating with those vessels at low data rates is made feasible by using a modified pulsed ladar system.

Although the Moon-to-Earth and Mars-to-Earth laser communications systems were canceled because of funding problems, nevertheless, the essence of the SPB is presented for the downlinks because it indicates the bounds affected by the very large distances that are involved and provides a vista into the future. Substituting for the laser communication system are microwave signals in the X- and K-Bands. However, useful calculations were performed by NASA agencies and components were developed for future missions of planetary laser communications.

Two major developments are considered for use between microsatellites and also nanosatellites. They are SALT and RRCS. The goal of the volume size and the weight of the SALT system, as currently planned, is 1 cm^3 and 1 g. These optimistic figures are likely to grow, but the concept appears doable and can become a major transceiver design for intersatellite communications and other communication platforms

The RRCS is a system in which a CW laser is transmitted from a mothership satellite to the receiver satellite to interact with a retromirror. However, before being reflected back to the mothership, the received CW is modulated by a signal source located at the receiver satellite, and then that beam is reflected back to the transmitting (mothership) satellite. This communications system is small in size, having few components, and it is highly reliable. Apart from application as an IFF function, it can become useful in a variety of other communications applications.

References

[1]   Data sources on high energy laser work are found at Naval Research Laboratories in Washington, D.C.; the Lawrence Livermore National Laboratories, in Livermore, CA; and Sandia Laboratories in Albuquerque, NM.

[2]   Jerlov, N. G. *Optical Oceanography,* Philadelphia: Elsevier Publishing, 1968.

[3]   Yura, H. T. "Propagation of Finite Cross-Section Laser Beams in Sea Water," *Applied Optics*, Vol. 12, No. 1, January 1973.

[4]   Yura, H. T. "Mutual Coherence of a Finite Cross Section Optical Beam Propagating in a Turbulent Medium," *Applied Optics,* Vol. 11, No. 6, June 1972.

[5]   Yura, H. T., "Small Scattering of Light by Ocean Water," *Applied Optics,* Vol. 10, No. 1, January 1971.

[6]   Blasé, P., "Laser Communications for a Lunar Base," The Artemis Society, ASI W 9700338rl.4, 2004.

[7]   Sweeting, M., "Micro/Nano Satellites—Brave New World," Royal Academy of Engineering, University of Surrey, October 2001.

[8]   Bister, K., "Steered Agile Laser Transceivers (SALT)," University of California, Berkeley CA 94720-1770; Proposal presentation, BAA99-25, "Steered Agile Beams."

[9]   Stoll, H. M., and E.M. Garmire, U.S. Patent Number 4,755,014, 1979.

[10]   Creamer, N. G. "Multiple Quantum Well Retro-Demodulators for Spacecraft-to-Spacecraft Laser Interrogation, Communication and Navigation," 15th Annual AIAA/USU Conference on Small Satellites, 2001.

[11]   Goetz, P. G., W. S. Rabinowich, R. J. Walters, S. C. Messenger, G. C. Gilbreath, R. Mahon, M. Ferraro, K. Ikossi-Anastasiou, and D. S. Katzer, "Effects of Proton Irradiation on InGaAs/AlGaAs Multiple Quantum Well Modulators," *Proc. IEEE,* AeroSense 2001 Paper no. 470 , March 2001, Big Sky, Montana.

About the Author

David G. Aviv has more than thirty years of experience in laser communication and radar development as an entrepreneur or lead technologist in small, medium, and large corporations. Along with radar development, Mr. Aviv has been involved in other active and passive sensors for ground, airborne, and space platforms, and specific perimeter defense applications. He has also developed application systems involving laser communications engineering covering cellular, ground-based mobile stations, and satellites, in different atmospheric and hostile environments.

David Aviv's initial work in laser space communication was as project engineer with U. S. Air Force Program 405B, which aimed at providing communication between synchronous satellites about 80,000 km apart. While the uplink/downlink was initially planned to be in microwaves, intensive studies were encouraged to develop a theoretical model for those links with laser beams. In particular, experiments initiated and supported by NASA and DoD were undertaken to test laser transceiver components and to measure atmospheric turbulence parameters. His work in the Strategic Defense Initiative involved high-intensity laser beams capable of remote energy deposition under laser-based command and control communications.

Mr. Aviv received his M.A. (mathematics) and M.S. in electrical engineering (communications) from Columbia University. He has taught extensively under the auspices of the Institute of Electrical and Electronics Engineers and has worked at RAND, The Aerospace Corporation, Rockwell, Lockheed, and currently at ARC Incorporated.

Index

Acquisition tracking and pointing (ATP), 9,
    15, 45–72
  accuracy pointing, 61
  baseline transreceiver with inertial sen-
    sors, 68–71
  beacon laser, 66–68
  block diagram of ATP on LEO and
    GEO, 51–54
  broad beam scanning, 54–55, 58–61
  fine scanning mirror (FSM), 46–47
  focal plane array (FPA), 46–47
  geosynchronous equatorial orbit
    (GEO), 45–72
  implementation of functions, 47–51
  inertial sensors on satellite, 68–71
  laser transceiver, 66–71
  low earth orbit (LEO) satellite, 45–72
  narrow beam scanning, 54–55
  point-ahead angle (PAA), 45, 47, 48
  signal-to-noise ratio (S/N), 62
  specific acquisition procedures, 54–61
  timing, 63–64
Adaptive optics subsystem (AOS), 10–11
  downlinks, 74–77, 95–97
  Fifth Generation Internet (5-GENIN)
    System, 123–39
  passive reflector configurations, 156
  uplinks, 87–88, 98–101
  weather issues, 104–5

AF Maui Optical Station (AMOS), 154
Antisubmarine warfare (ASW) system, 14,
    168–71
Articulating mirror system (AMS), 141,
    150–53
ATP. See Acquisition tracking and pointing
    (ATP)
Augmented tracking and acquisition system
    (ATAS), 154–55

Beer's Law, 105–10
Bit error rate (BER), 8, 9, 10–11, 21–25
  downlinks and calculation of, 77–82
  signal power budget (SPB) and, 21–25,
    37–42
Blue-green laser system design, 169
Border observation and monitoring,
    131–32

Clouds (weather issues). See Weather
Combined RF and laser telescope antenna,
    167–68
Constellation of low-altitude satellites,
    38–39
Continuous wave (CW), 166, 187
Crosslink, 7

Directivity, 3–4
Disaster Managements Interoperability Ser-
    vice (DMIS), 132

Downlinks, 7, 10, 73–84
  adaptive optics subsystem (AOS),
    74–77, 95–97
  analytic expressions, 77–82
  angle of arrival, 82–83
  BER calculation, 77–82
  Fifth Generation Internet System. *See*
    Fifth Generation Internet
    (5-GENIN) System
  measurements, 93–94
  microwaves, 7
  modulation transfer function (MTF),
    93–94
  OOK modulation, 77
  optical ground station (OGS), 76
  Reciprocity Theorem, 75
  reference signal for AOS, 95–97
  reference-adaptive optics and reciprocity
    (RAOR), 77
  satellite to ground station, 75–77
  spacecraft to earth-based optical station,
    73–84
  uplinks compared, 88

Echo project, 145–146

Far infrared (FIR), 6
Fifth Generation Internet (5-GENIN) Sys-
    tem, 12–13, 123–39, 167
  Aerostat endurance, 133
  airship vulnerability issues, 132–33
  AO telescope antenna, 129–30
  backbone satellites, 124–27, 135–38
  border observation and monitoring,
    131–32
  defined, 124
  Disaster Managements Interoperability
    Service (DMIS), 132
  GOES-East satellite, 127–29
  GOES-West satellite, 127–29
  high-value targets, 130–31
  Joint Biological Point Detection
    System Suite (JBPOSS),
    131–32
  miniaturized unmanned ground-based
    mobile (MUGM) systems, 123,
    129–30, 134–35
  mobile-type terminals, 123, 129–30,
    134–35
  RF power delivery, 136–38

synchronous backbone satellites,
    124–27, 135–38
weather issues, 126–28

Geosynchronous equatorial orbit (GEO),
    45–72
Geosynchronous operational environmental
    satellites (GOES), 127–28
Global Positioning System (GPS), 10
Ground based laser (GBL), 12, 156, 157

High-energy laser (HEL), 157, 161

Inertial sensors on satellite to detect vibra-
    tions, 68–71
Internet. *See* Fifth Generation Internet (5-
    GENIN) System
Interplanetary laser communication, 178–82

Joint Biological Point Detection System
    Suite (JBPOSS), 131–32

Ladar cross-section (LCS), 169
Ladar internal wave detection experiment
    (LIDEX), 169
Laser transceiver and ATP, 66–71
Long-wave infrared (LWIR), 6, 168
Low altitude satellite (LAS), 9–10
Low earth orbit (LEO) satellite, 45–72

Machine vision system (MVS), 182
Mars-to-Earth links
  interplanetary laser communication,
    179–80, 187
  Mars Reconnaissance Orbiter (MRO),
    35, 166
  pulse position modulation (PPM),
    34–35
Microsatellite, 182–83
Microwaves
  advantages of lasers over, 2–5
  combined laser and, 5–6
  downlinks, 7
  higher bandwidth of lasers *versus,* 4
  large directivity, 3–4
  narrow beamwidth and, 2
  privacy, 4–5
  uplink/downlink and laser crosslinks, 7
Mid-wave infrared (MWIR), 168
Miniaturized unmanned ground-based
    mobile (MUGM) systems, 13, 123,
    129–30, 134–35, 186

Mirrors. *See* Passive reflector configurations

MODTRAN system for laser signal penetration, 112, 121

Moon-to-Earth laser link, 180–82, 187

Multiple quantum well modulator (MQWM), 166, 186

Nanosatellite, 182–83

Narrow beamwidth, 2

National Aeronautics and Space Administration (NASA), 163

National Oceanic and Atmospheric Administration (NOAA), 12, 13

Near-wave infrared (NWIR), 6, 168

On-off keying (OOK), 8, 10, 40–41, 77

Optical ground station (OGS)
downlinks, 76
uplinks, 89
weather issues, 104, 121–22

Optical Westford, 158–59

Passive reflector configurations, 141–63
adaptive optics subsystem (AOS), 156
advantages, 142
AF Maui Optical Station (AMOS), 154
Apollo 15 (of National Aeronautics and Space Administration), 163
applications, 146–47
arbitrary reflective surface, 143
articulating mirror system (AMS), 141, 150–53
augmented tracking and acquisition system (ATAS), 154–55
data rate, 148–49
experiments using, 153–56
GPS used to locate reflective faces, 159–60
ground mirrors used to locate reflective faxes, 159–60
ground-based laser (GBL), 156, 157
high-energy laser (HEL), 157, 161
history, 145–46
modulation scheme, 148–49
nominal reference link, 147–50
Optical Westford, 158–59
Project Echo, 145–46
range, 148–49
Relay Mirror Experiment (RME), 13–14, 141, 153, 155

retrodirective mirrors used to locate reflective faxes, 159–60
signal power budget, 149–50

Point-ahead angle (PAA), 9, 45, 47, 48

Privacy comparison with microwave, 4–5

Pulse gated ninary modulation (PGBN), 8, 9, 19, 21–23, 42

Pulse position modulation (PPM), 32–35

Relay Mirror Experiment (RME), 13–14, 141, 153, 155

Retroreflective communication system (RRCS), 14, 166, 186, 187

Return link system (RLS), 166, 177–78

satellite-to-submarine communication, 176–77

Seawater turbulence, 172–74

Semiconductor Laser Intersatellite Link Experiment (SILEX), 118–20

Signal coupling efficiency (SCE), 11

Signal power budget (SPB)
azimuthal pointing error angle, 40–42
bit error rate (BER), 21–25, 37–42
calculation, 16–18
constellation of low-altitude satellites, 38–39
direct detection *versus* heterodyne detection, 25–35
elevation components of pointing error angle, 40–42
error analysis for PGBM, 22–23
expression of SPB due to vibrations, 35–39
heterodyne detection *versus* direct detection, 25–35
intersatellite link (ISL) application, 35, 37–38
intersatellite links, 15–42
loss of signal due to vibrations, 35–39
numerical example, 19–20
passive reflector configurations, 149–50
photoelectrons per bit and modulation scheme, BER, 21–25
pulse position modulation (PPM), 32–34
vibrations and their effect, 35–39

Signal-to-noise ratio (S/N), 62

Space-based detectibility and identification of submersibles (SBDIS), 14, 165, 168–71

SPB. *See* Signal power budget (SPB)
Steered agile laser transreceiver (SALT), 14, 166, 183–87, 187
Submarine laser communication to satellite (SLCSAT), 166, 171–74
Synthetic aperture radar (SAR), 6
Synthetic sodium laser beacon, 96–97

Terrestrial links
  weather issues, 103–22
Transformation Communication Architecture (TCA), 14, 167
Transformational satellite (ISAT) System, 14

Uplinks, 87–101
  adaptive optics subsystem (AOS), 87–88, 98–101
  aperture averaging subsystem (AAS), 92
  associated atmospheric turbulence, 92–93
  coherence length, 92–93
  downlinks compared, 88
  Fifth Generation Internet System. *See* Fifth Generation Internet (5-GENIN) System
  loss of signal when AOS not used, 98–101
  measurements, 93–94
  microwaves, 7
  modulation transfer function (MTF), 93–94
  optical ground station (OGS), 89
  oriented mirror
  reference downlink signal for AOS, 95–97

  reference laser, 97–98
  signal coupling efficiency, 89–92
  synthetic sodium laser beacon, 96–97
  zenith angle, 89–92

Weather
  absorption and scattering in atmosphere, 105–10
  aerosol scattering, 105–10
  AOS, 104–5
  attenuation due to absorption and scattering, 110–12
  Beer's Law, 105–10
  calculation of atmospheric turbulence parameters, 104–5
  dry weather locations in southwest region, 113–18
  Fifth Generation Internet (5-GENIN) System, 126–28
  MODTRAN system for laser signal penetration, 112, 121
  optical ground station (OGS), 104, 121–22
  penetrating low clouds surrounding ground station or airstrip, 120–21
  scatter and absorption constants, 105–10
  scattering coefficient for cloudburst condition, 110
  Semiconductor Laser Intersatellite Link Experiment (SILEX), 118–20
  terrestrial links and, 103–22
  testing laser communications along terrestrial links, 118–20
  vaporizing hole through cloud, 120–21
  weather avoidance system (WAS), 12–13, 103–104, 112–18

Recent Titles in the Artech House
Space Technology and Applications Series

Bruce R. Elbert, Series Editor

Business Strategies for Satellite Systems, D. K. Sachdev

Gigahertz and Terahertz Technologies for Broadband Communications, Terry Edwards

Ground Segment and Earth Station Handbook, Bruce R. Elbert

Introduction to Satellite Communication, Second Edition, Bruce R. Elbert

Laser Space Communications, David G. Aviv

Low Earth Orbital Satellites for Personal Communication Networks, Abbas Jamalipour

Mobile Satellite Communications, Shingo Ohmori, et al.

Satellite Broadcast Systems Engineering, Jorge Matos Gómez

The Satellite Communication Applications Handbook, Second Edition, Bruce R. Elbert

Satellite Communications Fundamentals, Jules E. Kadish and Thomas W. R. East

The Satellite Communication Ground Segment and Earth Station Handbook, Bruce Elbert

Understanding GPS, Elliott D. Kaplan, editor

For further information on these and other Artech House titles, including previously considered out-of-print books now available through our In-Print-Forever® (IPF®) program, contact:

Artech House   Artech House
685 Canton Street   46 Gillingham Street
Norwood, MA 02062   London SW1V 1AH UK
Phone: 781-769-9750   Phone: +44 (0)171-973-8077
Fax: 781-769-6334   Fax: +44 (0)171-630-0166
e-mail: artech@artechhouse.com   e-mail: artech-uk@artechhouse.com

Find us on the World Wide Web at: www.artechhouse.com